Metallic Contaminants
and Human Health

ENVIRONMENTAL SCIENCES

An Interdisciplinary Monograph Series

EDITORS

DOUGLAS H. K. LEE
National Institute of
Environmental Health Sciences
Research Triangle Park
North Carolina

E. WENDELL HEWSON
Department of
Atmospheric Science
Oregon State University
Corvallis, Oregon

DANIEL OKUN
University of North Carolina
Department of Environmental
Sciences and Engineering
Chapel Hill, North Carolina

ARTHUR C. STERN, editor, AIR POLLUTION, Second Edition.
In three volumes, 1968

L. FISHBEIN, W. G. FLAMM, and H. L. FALK, CHEMICAL MUTAGENS: Environmental Effects on Biological Systems, 1970

DOUGLAS H. K. LEE and DAVID MINARD, editors, PHYSIOLOGY, ENVIRONMENT, AND MAN, 1970

KARL D. KRYTER, THE EFFECTS OF NOISE ON MAN, 1970

R. E. MUNN, BIOMETEOROLOGICAL METHODS, 1970

M. M. KEY, L. E. KERR, and M. BUNDY, PULMONARY REACTIONS TO COAL DUST: "A Review of U. S. Experience," 1971

DOUGLAS H. K. LEE, editor, METALLIC CONTAMINANTS AND HUMAN HEALTH, 1972

DOUGLAS H. K. LEE, editor, ENVIRONMENTAL FACTORS IN RESPIRATORY DISEASE, 1972

H. ELDON SUTTON and MAUREEN I. HARRIS, editors, MUTAGENIC EFFECTS OF ENVIRONMENTAL CONTAMINANTS, 1972

DOUGLAS H. K. LEE and PAUL KOTIN, editors, MULTIPLE FACTORS IN THE CAUSATION OF ENVIRONMENTALLY INDUCED DISEASE, 1972

In preparation

MOHAMED K. YOUSEF, STEVEN M. HORVATH, and ROBERT W. BULLARD, PHYSIOLOGICAL ADAPTATIONS: Desert and Mountain

Fogarty International Center Proceedings No. 9

Metallic Contaminants and Human Health

Scientific Editor
Douglas H. K. Lee
NATIONAL INSTITUTE OF
ENVIRONMENTAL HEALTH SCIENCES
RESEARCH TRIANGLE PARK, NORTH CAROLINA

Sponsored by
National Institute of Environmental Health Sciences
Research Triangle Park, North Carolina
and
John E. Fogarty International Center
National Institutes of Health
Bethesda, Maryland

Academic Press
New York and London 1972

ACADEMIC PRESS, INC.
111 Fifth Avenue, New York, New York 10003

United Kingdom Edition published by
ACADEMIC PRESS, INC. (LONDON) LTD.
24/28 Oval Road, London NW1 7DD

LIBRARY OF CONGRESS CATALOG CARD NUMBER: 78-187864

CONTENTS

CHAPTER 1. INTRODUCTION: CONCEPTS OF ENVIRONMENTAL TOXICOLOGY

Emil A. Pfitzer

PART I. MAJOR ENVIRONMENTAL CONTAMINANTS

CHAPTER 2. MERCURY

Leonard J. Goldwater and Thomas W. Clarkson

CHAPTER 3. LEAD

Robert A. Goyer and J. Julian Chisolm

CHAPTER 4. CADMIUM
D. W. Fassett

PART II. OTHER CONTAMINANTS

CHAPTER 5. BERYLLIUM
Lloyd B. Tepper

CHAPTER 6. FIVE OF POTENTIAL SIGNIFICANCE
Ralph G. Smith

CHAPTER 7. FLUORIDES
Harold C. Hodge and Frank A. Smith

PART III. ADDITIONAL CONSIDERATIONS

CHAPTER 8. NUTRITIONAL ASPECTS OF METALS
M. R. Spivey Fox

CHAPTER 9. CASE FINDING: USES AND LIMITATIONS
J. Julian Chisolm

CHAPTER 10. ANALYTICAL CONSIDERATIONS
Lloyd B. Tepper

CONTENTS

ORGANIZING PANEL

Thomas W. Clarkson, Associate Professor of Pharmacology, School of Medicine and Dentistry, University of Rochester, Rochester, New York 14620

Peter G. Condliffe, Chief, Conference and Seminar Program Branch, Fogarty International Center, National Institutes of Health, Bethesda, Maryland 20014

Robert A. Goyer, Department of Pathology, University of North Carolina, Chapel Hill, North Carolina 27515

Robert Horton, Medical Advisor, National Environmental Research Center, Environmental Protection Agency, Research Triangle Park, North Carolina 27711

Douglas H. K. Lee, Associate Director, National Institute of Environmental Health Sciences, Research Triangle Park, North Carolina 27709

Lloyd B. Tepper, Associate Director, Department of Environmental Health, University of Cincinnati, College of Medicine, Cincinnati, Ohio 45219

CONTRIBUTORS

J. Julian Chisolm, Department of Pediatrics, Baltimore City Hospital, Baltimore, Maryland 21224

D. W. Fassett, Laboratory of Industrial Medicine, Eastman Kodak Company, Rochester, New York 14650

M. R. Spivey Fox, Division of Nutrition, Food and Drug Administration, Department of Health, Education, and Welfare, Washington, D. C. 20204

Leonard J. Goldwater, Department of Community Health Sciences, Duke University Medical Center, Durham, North Carolina 27706

Harold C. Hodge, Department of Pharmacology, University of California School of Medicine, San Francisco, California 94122

Emil A. Pfitzer, Department of Environmental Health, University of Cincinnati, Cincinnati, Ohio 45219

Frank A. Smith, Department of Radiation Biology and Biophysics, University of Rochester School of Medicine and Dentistry, Rochester, New York 14620

Ralph G. Smith, Department of Industrial Health, University of Michigan School of Public Health, Ann Arbor, Michigan 48104

EDITORIAL COMMENT

As governmental and public interest in environmental quality increases, and as conservation or remedial programs are planned, a number of scientists are called upon for advice or decision in areas involving fields beyond their own personal expertise. Program managers, experienced in administration, also find themselves in need of information on technical matters that fall within their jurisdiction.

The announcement that a United Nations Conference on human environment is to be held in Stockholm in June, 1972 pointed out the fact that these needs are worldwide. It stimulated the John H. Fogarty International Center of the National Institutes of Health to investigate how these needs might be met. In conjunction with the National Institute of Environmental Health Sciences, also part of the National Institutes of Health, a decision was made to prepare four books on the aspects of environmental health for which suitable résumés were not readily available.

Metallic contaminants and human health was the topic selected for the initial effort. A small panel of experts in the field was asked to delimit the scope that should be followed, to indicate the specific topics within that scope, and to suggest other experts who could contribute to the volume. Contributors brought draft papers to a three-day workshop where the drafts were thoroughly discussed, amendments suggested, and integration between chapters developed. Extensive editing followed.

It has not been easy to preserve a balance between simplification for the nonspecialist and adequacy as viewed by the expert. The text now appearing has been checked by the contributors, but I must accept responsibility for any undue selectivity that may have occurred, as well as for errors of omission or commission. I hope, however, that the text will help those who need to know the state of current knowledge on the health significance of metallic contaminants but who do not have the time to pursue the detailed literature or to seek a compilation directed to their special needs.

Douglas H. K. Lee

CONCENTRATION UNITS AND
CONVERSION FACTORS

Metric and proportional units are used somewhat indiscriminately for indicating the concentration of toxic agents. In this volume the units preferred by the individual contributor have been retained. The reader who wishes to make comparisons between concentrations expressed in different units will find the following data useful.

In solid and liquid mixtures, proportional units refer to weights and are easily converted to metric equivalents:

$$1 \text{ part per million (ppm)} = 1 \text{ mg/kg}$$
$$1 \text{ part per billion (ppb)} = 1 \text{ } \mu\text{g/kg}$$
$$(1 \text{ kg of liquid of density } 1 = 1 \text{ li})$$

In gaseous mixtures, however, proportional units refer to volumes, so that conversion varies with the molecular weight of the dispersed substance, temperature, and barometric pressure. The conversion formula is

1 part per million by volume
at 25°C and 760 mm Hg pressure = 0.041 (molecular
weight) mg/m^3

Note: The factor 0.041 represents $273/(298 \times 22.4)$

Particulate matter in air is often expressed in terms of millions of particles per cubic foot or per cubic meter. To convert from cubic foot to cubic meter, multiply by 35.3. (This also gives particles per cubic centimeter.)

CHAPTER 1. INTRODUCTION: CONCEPTS OF ENVIRONMENTAL TOXICOLOGY

EMIL A. PFITZER, Department of Environmental Health, University of Cincinnati, Cincinnati, Ohio

Concerned citizens and scientists alike face the burning issue of the need for action for the elimination, control, and clean-up of contamination which has been created by man's technological activities. The concerned citizen wishes for action now to clean up the environment and to establish controls for the prevention of environment-induced disease. The concerned scientist must, in addition, obtain the facts which form the basis for judgments about how much clean-up and how much prevention. Final decisions about control and prevention of environment-induced diseases will usually be made through the political process. Legislators will make these decisions on demands from the citizenry; they should reflect consultation with sociologists, economists, lawyers, and engineers, as well as with health scientists. Environmental toxicologists, along with other health scientists, must present the biological facts as information for the decision-maker. They must also communicate the general underlying concepts to the citizenry, to the public authorities, and to their fellow scientists.

Definition

Environmental toxicology is the study of the unwanted effects of chemical environmental agents on living things. This is a definition which is simple in words but broad in concept, and thus requires elaboration.

When Is an Effect "Unwanted"?

One might first ask the question, "unwanted by whom?", because some of us, on occasion, deliberately seek the effects of tobacco smoke or alcoholic beverages, while others abhor such effects with sincere passion. The fact is that most of us will accept some degree of some effects from chemical agents in our environment; our environment including air, water, food, drugs, and cosmetics.

While we might never get universal agreement about "wanted" and "unwanted" effects, we probably can come the closest on these two points: (1) the effect should not be adverse to one's health, and (2) one should have free choice to accept or not to accept an exposure which produces a given effect. Additional extremely important points which always elicit much debate include the unwanted effects which are classified as nuisances rather than effects on health, and the unwanted effects which induce ecological imbalances with potentially far-reaching consequences.

"Wanted" vs "Tolerable" Effects

It is easy to list some wanted effects of chemical agents deliberately introduced into our environment. We demand the satisfaction to our taste buds which comes from the multitude of chemical flavoring agents added to our food and beverages. Many of us religiously take and give to our children an assorted chemical mixture of vitamins and minerals in order to enhance growth and vitality.

The concept of tolerable effects implies that basically the effects are unwanted but that we can accept them if there is good reason. This is the big gray area between wanted and unwanted effects. The easiest example to accept is that of the side effects of therapeutic drugs which are basically unwanted, but tolerated in order to get the benefit of the drug in treating a disease or bodily malfunction. In addition we tolerate the addition of chemical preservatives to much of our food supply, because if we did not the food would spoil more readily and malnutrition, already too common, would increase. There could follow a long list of effects that we tolerate in order to have the benefits of nutrition,

2

therapeutic medicine or comforts brought by technological products such as automobiles, electric power, metal and plastic containers and numerous others.

This concept of tolerable effects is implicit in the age-old consideration of judgments between benefits and risks. The role of the environmental toxicologist is to define and to quantitate the risks (as unwanted effects) so that societies may intelligently weigh the effects against related benefits and decide what shall be tolerated.

What Is an Adverse Health Effect?

This may sound like a very naive question and, indeed, the toxicologist has no difficulty in treating death, tissue necrosis, and gross biochemical or functional changes as adverse effects. The difficulty arises in the assessment of an effect as not being adverse, but simply as being a normal, physiologically acceptable change, or even as a desirable defensive action by the body. An adverse, or abnormal, effect is often defined in terms of a measurement which is outside of the normal range. The normal range, in turn, is usually defined as that range of measured values observed in a group of presumably healthy individuals. These measured values are often expressed in statistical terms as an average value, and a range representing 95 percent confidence limits of the average. An individual with a measured value outside of the above range may be either "abnormal" in fact or one of that small group of "normal" individuals who have extreme values. This dilemma serves to emphasize that difficult gray area for distinction between "normal" and "abnormal", and its application to the definition of an adverse health effect.

The conditions under which individuals are measured must be carefully standardized. For example a heart rate of 120 beats per minute may be "normal" during exercise, but "abnormal" when the individual has recovered and is at rest. The elevated heart rate may be considered as an acceptable normal value during periods of exposure to exercise. Similarly an industrial worker may be exposed to a specific chemical and have an elevated concentration of that

3

chemical in his body, but be healthy by all clinical criteria. The range in the body for that chemical in a group of industrial workers may be higher than for non-industrial workers, but "higher" in this case would not imply an "adverse health effect". The normal range for the lesser exposed individuals is sometimes called the "natural" level, on the presumption that their exposure is that found only "in nature" and not subject to man's manufacturing processes.

The toxicologist must use the most sensitive tools for measurement and the most appropriate statistical procedures for interpreting the significance of small differences between measurements. He must make a judgment as to whether the measurement reflects an adverse effect on health or a normal physiologically acceptable change. This is the most crucial role of the environmental toxicologist, because society rightfully insists on knowing about the significance of the smallest risks. It is easy for society to decide to eliminate the high risks, but it is difficult to decide when risks are small enough to be tolerated; the environmental toxicologist must quantitate these small risks for adverse health effects in a realistic and accurate manner.

Selective Localization vs Selective Action

When one utilizes an increased level of a chemical in the body as an indicator of an adverse health effect, it is implied that there is a significant correlation between the concentration of the chemical and adverse effects. This correlation must be established by biological data, however, since it is known that locations in the body which selectively contain the highest concentrations are not necessarily the locations where the most significant and selective toxic action takes place. For example, lead tends to localize selectively in bones at sites where it is primarily biologically inactive. Mercury may be present in the kidneys at elevated levels for years without any evidence of selective toxic effect.

Exposure to Chemical Environmental Agents

Man has from his earliest history been engaged in the process of taking materials from the earth, the water and the air and then using them with increasing technological ingenuity to make new chemical agents, eventually returning them in some form to the earth, water, and air. This process provides a broad array of man-made exposures to numerous chemical environmental agents which might be contrasted with those of primitive man. That the effects of man-made exposures have sometimes been beneficial, even life-saving, as well as sometimes detrimental to health, is indisputable. The environmental toxicologist must attempt to quantitate the conditions of exposure that produce unwanted effects, a task which is monumental when one considers the vast number of chemical agents in the environment, the variable circumstances and duration of exposure and the complexity of health effects which may result from interactions of simultaneous exposures to multiple agents.

What Happens in Between "Ashes-to-Ashes"?

The concept that chemical agents in the environment are transported through a cyclic process is not altogether reassuring when one considers the time necessary for portions of some cycles to take place. These time lags may permit build-up of specific chemical agents to concentrations never before experienced. One would like to know the complete natural history of occurrence, transport, transformation, accumulation, and degradation for all chemical forms involved in the "ashes-to-ashes" cycle.

That increased population, technological advances, and improved -- perhaps more artificial -- living conditions have combined to make the cycle more complex is obvious. Experiences with DDT and mercury have dramatized the fact that unexpected events related to the "natural history" of these chemicals can lead to unwanted effects for man and other living things. Can environmental toxicology, combined with ecological considerations, provide information to prevent or minimize the hazards due to such events?

The need for biodegradable detergents and pesticides has been emphasized in recent years. Special attention must be paid to those chemical agents which are less readily degraded to "natural" products. Metallic contaminants are unfortunately unique in that they do not degrade except by nuclear fission. Therefore metals, as such, revolve through the cycle by changing only their physical form or chemical attachment. Thus lead sulfide ore may be converted to lead metal, which may be alloyed with sodium and converted to tetraethyl lead, and this may be converted in the combustion engine to lead chlorobromide or lead carbonate. Our major concern with metallic contaminants, then, is their redistribution from fairly remote deposits to populated areas where they may accumulate and become a hazard.

Interactions Between Living Things

Most environmental toxicologists collect data on laboratory animals which are subsequently extrapolated for application to man. Occasionally a chemical may affect laboratory animals but not man, as appears to be the case with some chemicals which produce cancer in laboratory animals. Extrapolations from animals to man, therefore, must be made with judgment. A knowledge of the similarities and differences between man and laboratory animals is essential. Recent expressions of concern have emphasized that the effects of chemical environmental agents on non-human life, specifically wildlife, cannot be ignored, and that these effects may not always be independent of man's health.

These concerns are in part those of the conservationist interested in maintaining the living species present today. They certainly cannot be ignored, even where the resulting imbalance of nature is not of obvious import to man's health. Consideration of health, as being a state of total well-being, can legitimately include esthetic qualities as an appropriate item in the "benefits" column. Of more alarming concern are those prophecies about effects of chemical environmental agents which may lead to such drastic imbalances as to imperil the earth. By and large such prophecies of doom are counterbalanced by reasoned scientific logic which belies the prophecy. Nevertheless it would be imprudent to ignore the consideration that man's disrespect for certain other living things could possibly lead to irreversible consequences which would be destructive for mankind.

The environmental toxicologist traditionally measures only unwanted effects on plant and animal life. However, his observations on the environmental conditions which cause effects, and the evidence that he develops on distribution and storage sites within living things, may provide important clues about unexpected effects of interaction between living things.

Concept of Toxicity and Hazard

Toxicologists have found it useful to give special meanings to the words toxicity and hazard. Semantics apart, the toxicologist will generally define toxicity as the inherent property of a chemical to produce an unwanted effect when the chemical has reached a sufficient concentration at a certain site in the body, and hazard as simply the probability that such will occur. This usage recognizes the fact that an observed unwanted effect in a living thing will usually be the result of a chemical's ability to enter an organism, be transported, pass through various membranes, by-pass various defense mechanisms and finally reach a specific site in sufficient concentration for reaction.

This concept is important in that it helps us to understand why in one case a chemical placed on the skin may have difficulty passing through the membranes of the skin and thus not cause any observable effects, while the same or smaller quantity of the same chemical inhaled into the lungs may be absorbed across the alveolar membrane into the blood and be rapidly carried to a site where a major unwanted effect occurs. The toxicologist would say that this toxic chemical presented a small hazard when only in contact with the skin, but a high hazard when inhaled. The same concept helps us to understand why inhaled particles of relatively large size, that "fall out" onto the airways of the lungs and are cleared by the defense mechanism of mucous elimination, appear harmless, while the same chemical particles in relatively small size, that can be carried deep into the lungs where they may react with sensitive tissue, or be dissolved and absorbed into the body, exert a toxic effect.

When comparing chemical environmental agents one must recognize that the measurement of the overall unwanted effect reflects the relative toxic hazards rather than just relative toxicities. Relative toxicity would apply only to specific conditions of exposure; a different set of exposure conditions, posing a possibly different set of barriers to the chemicals, may well give different comparative results. In addition, variations from individual to individual called host factors, such as amount of body fat, state of health, etc., can be responsible for large variations in the observed toxic hazard.

Concept of Non-toxicity -- A Fallacy

Of equal importance to the recognition that the method of entry and the form of the chemical agent are important, is the concept that every chemical agent is inherently toxic. Thus any chemical agent given in large enough quantity, by the appropriate route of administration, and in the proper physical form, will produce an unwanted effect. The popular term of "non-toxic" as used on some labels, really means "low toxic hazard by the oral route".

If this concept strikes one as strange, it may be recalled that pure oxygen (as opposed to the normal 20% concentration in air) has a toxic effect, and that too much water in the wrong place (the lungs) kills many people every year. An understanding that every chemical agent has some degree of inherent potential for toxicity will help one keep alert for unexpected results when the chemical agent is used in a new and different manner.

Concept of a Safe Level of Exposure

Just as the fact that every chemical is toxic is a basic principle of toxicology, so are the concepts that most chemicals can be handled safely, and that the body can tolerate some level of exposure without any observable unwanted effect. The tolerable level may be so low that the only practical attitude is to ban its use (one form of "zero tolerance"), but the concept is valid.

These concepts are basic to considerations of evaluation of safety and to the setting of exposure limits. The environmental toxicologist has the responsibility of estimating the quantity of the chemical agent which will not produce an unwanted effect under stated conditions of exposure.

There are chemical environmental agents, specifically the carcinogens, which are considered to be capable of a non-reversible reaction at some sites leading to tumor growth, and possible death. This is termed by some toxicologists as a no-threshold interaction, in that theoretically one molecule of a chemical will initiate a reaction which can lead to death. In this sense the body, indeed, has no threshold, or "zero-tolerance", for the one molecule which reached the critical site of reaction. However, in order for this one molecule to reach the critical site it may be necessary for very large numbers of molecules to enter the body. Such a situation, when it can be estimated in quantitative terms of concentration or time, defines an actual threshold relative to practical exposure conditions, even though an individual molecular interaction, in itself, may lead to an adverse effect.

Thus one basis for establishing a safe level of exposure may be that finite exposure to a chemical agent at which the likelihood of a molecule getting to a critical site is so remote as to be immeasurable. More acceptable bases for establishing safe levels of exposure are: (1) that the body can easily detoxify the dose through normal physiological mechanisms; (2) that the lesions produced are easily repaired by body mechanisms with no resultant unwanted effect; and (3) that the lesions produced are so inconsequential that they do not influence the reserve capacity for the maintenance of normal bodily functions.

Concept of Absolute Safety -- A Non Sequitur

Although the word "safe" has been used in the foregoing paragraphs, there really is no such thing as a guarantee of absolute safety, for the simple reason that the only proof is in the end result. Thus to say that all humans may be exposed safely for thirty years to X concentration of chemical A is proven absolutely true only after all humans have tolerated this exposure, and even then the questions remain: what about sixty or seventy or eighty years of exposure; what about the next generation, and the next, and the next? The questions and the time required for absolute proof are infinite.

What do we do, then? We use scientific judgment (sometimes very suspect) to make a prediction which we claim to be beyond all reasonable doubt (but not absolute). If ten or one hundred humans tolerate a given exposure to a chemical environmental agent, it is reasonable to assume that one thousand or one hundred thousand might also tolerate it, but it is not proof. In fact, there are instances in which no sensitive individuals are observed until the numbers of exposed people have reached the thousands. Toxicologists have learned that a new product should be used only by gradually increasing numbers of people, in order to eliminate unwitting exposure of large numbers of sensitive individuals.

Some chemical agents deliberately added to our environment provide benefits that far exceed the adverse health effect in only one of a million people. On the other hand, albeit an unpopular concept, many of us accept appreciable health risks every day on the assumption that the benefits to us outweigh the potential risks.

There is no way to avoid the facts of life that judgments, subject to error, must be made, that risks are taken and that evaluations of safety are relative, not absolute. Additionally, it is imperative that we carefully consider whether, in removing one potential risk, we might introduce alternatives with even more serious risks.

Concept of the Negligible Quantity

When a certain chemical will be present in only very small quantities, in a physical form which is not considered as highly hazardous, and by biochemical analogy is not suspect of being highly toxic, the judgment is sometimes made that it has a high probability of being used safely, although there are no direct data to that effect. In making this judgment, the environmental toxicologist has invoked the concept of the negligible quantity.

This concept has great practical significance in that large numbers of chemicals, including many natural flavors, are used in such small total quantity (worldwide) that the costs for toxicity testing would far exceed any reasonable part of the market value of the chemical. Of course one cannot justify allowing a hazardous chemical to be added to our environment without toxicity evaluation simply because one cannot afford the tests. The use of the concept of negligible quantity must only be used after open review and critique by experts and realistic examination of benefit-risk considerations.

Earlier in the development of environmental toxicology the concept of "zero tolerance" was used. This concept meant that one did not have sufficient data to establish a safe level and thus one chose to have none present. In theory this sounds logical, and in practice it was workable until analytical chemists laid bare our ignorance of the presence in our environment of many chemicals in extremely small quantities, whose presence had not been previously suspected. In the opinion of many investigators the zero tolerance concept is not scientifically sound, and should be replaced by the concept of finite tolerable levels and the concept of the negligible quantity.

It should be recalled that not every chemical environmental agent has zero as its ultimate desirable quantity. Nutritional studies have found that chemicals such as metals, which may be hazardous at high levels, may be advantageous, even necessary, in small quantities. These metals are called essential trace metals and include copper, zinc, cobalt, and chromium. In general, the nutritionist rather than the environmental toxicologist is concerned with these problems, although there is the theoretical possibility that some other chemical agents may interact and create artificial deficiencies of the essential metals.

Interrelationships: Industry   and   the
Community

Chemical environmental agents arise from natural
processes, from normal activities of a community
(sewage, fuel use in homes and vehicles, etc.) and
from industrial and commercial activities. The
environmental toxicologist is often hard put to
distinguish one source from another when he evaluates
the exposure of a given population under study. In
general, it is necessary for the environmental
engineer to survey the conditions in great detail.
Nevertheless, there are circumstances where the
source and the exposed population (within industry or
within the community) are well circumscribed. Such
locations become extremely important areas for joint
studies by engineers, physicians, and toxicologists
in order to evaluate any causal relationship between
unwanted effects and chemical environmental agents.

The population of industrial workers is of
particular importance in that they will generally
have a higher exposure to specific agents than do
people in the community, and will encounter new
materials at an early stage of development. Thus
safety evaluation for industrial workers can provide
the basis for safety evaluation of the parallel
population in the general community, that is,
reasonably healthy adults. It is important to
recognize that exposure in the community is
continuous (24 hours per day) as opposed to the
typical 40-hour work week. In addition, many people
in the general community are not reasonably healthy
adults and thus community studies of populations
including the chronically ill, children, and pregnant
women must be continued in order to find sensitive
groups or individuals.

Summary

Figure 1 may serve as a simplified hypothetical
expression of the interrelationship between exposure
to chemical environmental agents and unwanted effects
in living things. The environmental toxicologist
must examine our present knowledge about each
chemical environmental agent and provide the
information which can be used by society-at-large to
control and improve the environment for the
well-being of living things.

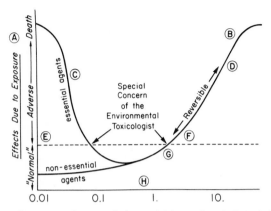

Fig. 1. Significant Activities to Establish the Relationship Between Exposure to Chemical Environmental Agents and the Effects Due to the Exposure:

A. The environmental toxicologist with assistance for all of the biological and clinical sciences attempts to describe and quantitate the sub-lethal unwanted effects observed in living things.

B. The coroner's toxicologist investigates deaths due to chemical environmental agents; the experimental toxicologist studies cause of death in animals; the clinical toxicologist handles cases of accidental poisoning.

C. The environmental toxicologist and nutritionist investigate deficiency states of essential chemical agents in human, animal and plant populations

D. The experimental toxicologist looks for changes in physiological function, biochemical and pathological changes, and mechanism of non-lethal effects; the industrial physician studies workers who have been impaired by environmental exposures.

E. The environmental toxicologist looks for the most sensitive change which can be called an unwanted effect; he also measures variation within normal limits, attempts to differentiate between physiological changes which are normal, even desirable, and those which may be , detrimental and, thus unwanted.

F. The environmental toxicologist searches for the most sensitive and specific changes that can be established as unwanted effects following long-term, low-level exposures; the epidemiologist searches among population groups for causal relationships between unwanted effects and exposure to a chemical environmental agent.

G. The environmental toxicologist extrapolates from animal data and human experience to estimate safe exposure levels; he measures levels of the chemical environmental agents in normal people to establish the body burden of the chemical agents.

H. The environmental toxicologist and the environmental engineer assess exposure conditions with regard to concentration, exposure time, repetition of exposures, physical and chemical form of the chemical agent, demographic description of the exposed population, etc.

GENERAL REFERENCES (used but not cited)

Frawley, J. P. (1968). A reasoned approach to regulation based on toxicologic considerations. Food Drug. Cosmet. Law J. 23: 260.

Hatch, T. F. (1968). Significant dimensions of the dose-response relationship. Arch. Env. Health 16: 571.

Oser, B. L. (1970). Reflections on some scientific problems in regulatory compliance. Assoc. Food & Drug Officials U. S., Quart. Bull. 34: 129.

Smyth, H. F. (1958). Safety evaluation procedures and interpretations. Food Technology 12: 17.

Smyth, H. F. (1963). Industrial hygiene in retrospect and prospect - toxicological aspects. Amer. Ind. Hyg. Assoc. J. 24: 222.

Smyth, H. F. (1967). Sufficient challenge. Food Cosmet. Toxicol. 5: 51.

Smyth, H. F. (1967). Defining potential health hazards emerging from changing technology. Amer. Ind. Hyg. Assoc. J. 28: 408.

Stokinger, H. E. (1969). Cummings Memorial Lecture: The spectre of today's environmental pollution - USA brand: new perspectives from an old scout. Amer. Ind. Hyg. Assoc. J. 30: 195.

Weisburger, J. H. and Weisburger, E. K. (1968). Food additives and chemical carcinogens: On the concept of zero tolerance. Cosmet. Toxicol. 6: 235.

PART I

MAJOR ENVIRONMENTAL CONTAMINANTS

CHAPTER 2. MERCURY

LEONARD J. GOLDWATER, Department of Community Health
Sciences, Duke University Medical Center,
Durham, North Carolina

THOMAS W. CLARKSON, School of Medicine and Dentistry,
University of Rochester, Rochester, New York

ECOLOGY

SOURCES

Geological

Mercury is one of the less abundant elements in
the earth's crust, being 74th in a list of 90,
comprising, according to one estimate (1), about 2.7
x 10^{-6} percent. Greater than trace amounts are found
in at least 30 ores, but from only one, the sulfide
cinnabar, does the concentration justify commercial
extraction. Some cinnabar ores are so rich that
droplets of the free metal are present in the native
state, and even exude to form small pools in the
mines.

Mercury mines are known in at least 40
countries, distributed in all continents except
Antarctica. Cinnabar was used in China at least as
early as 1100 B.C., and mines of comparable age are
believed to have been worked in Asia Minor and on the
island of Syra in the Cyclades. Mined cinnabar
dating back at least to 500 B.C. has been found in
Peru. Mines at Almaden in Spain are the richest
historical source, with records back to at least the
fourth century B.C. Important mines and recent
production figures are given in Tables 1 and 2.

TABLE 1 -- DISTRIBUTION OF MERCURY IN THE HYDROSPHERE

Country	Water Body	Year	Concentration (μg/l l)	Reference
France	Sea	1799	Present	Proust (2)
Germany	North Sea	1934	0.03	Stock (3)
"	Fresh water	"	0.01 - 0.10	"
"	Rain water	"	0.05 - 0.48	"
U.S.S.R.	Rivers, lakes, seas	1962	0.09 - 0.28	Aydin'yan (4)
Japan	Rain water	1964	0.6	Fujimura (5)
"	Ditches	"	4 - 100	"

Industrial

Mining and transportation of cinnabar, and extraction of the metal from the ore, reached significant proportions during the period 500 B.C. to 500 A.D. New compounds, new uses, and new opportunities for ecological cycling have progressed with increasing speed in the present century. Chlorides and oxides of mercury were made and known, particularly by the alchemists, before 1000 A.D. These and other inorganic compounds and mixtures of mercury were widely used by physicians and chemists throughout the middle ages. The synthesis of organic compounds of mercury by von Hoffmann in 1843, and by Frankland in 1850, led to the eventual creation of hundreds of new mercurials. Even a short list of present-day uses of the metal and its compounds would indicate their ubiquity.

CYCLES

For millions of years mercury cycled between the lithosphere, hydrosphere, and atmosphere in accordance with natural laws, as indicated in Fig. 1. In the last thousand years man-made perturbations have been imposed with increasing frequency and amplitude. In removing the element from natural ore sinks, developing an ever-widening range of compounds, and distributing the metal and its compounds over the face of the earth, man has vastly increased the amount in circulation at any one time. Only slowly, after many transitions and through

TABLE 2 -- MERCURY IN SELECTED STREAMS OF USA, 1970

U. S. Geological Survey (6)

Stream	Location	Date	Concentration (μg/l1)
Pemigewasset	Woodstock, N.H.	June 8	3.1
Merrimack	Above Lowell, Mass.	June 8	1.2
Hudson	Dowstream from Poughkeepsie, N.Y.	Apr. 7	.1
Delaware	Port Jervis, N.Y.	Apr. 23	<.1
Susquehanna	Johnson City, N.Y.	July 6	.1
Chemung	Near Wellsburg, N.Y.	July 6	.2
North Branch Potomac	Near Barnum, W. Va.	June 3	1.2
Pascagoula	Merrill, Miss.	June 9	3.0
Ohio	Near Grand Chain, Ill.	June 26	.1
Maumee	Antwerp, Ohio	June 10	6.0
Oswegatchie	Gouverneur, N.Y.	June 16	1.2
Rainy	International Falls, Minn.	May 14	<0.1
Wisconsin	Near Nekousa, Wis.	June 10	2.4
Bighorn	Kane, Wyo.	June 30	<.1
Floyd	Sioux City, Iowa	June 9	.2
North Platte	Near Casper, Wyo.	June 23	.1
Missouri	Hermann, Mo.	June 24	-
Mississippi	Near Hickman, Ky.	June 25	<.1
St. Francis	Marked Tree, Ark.	June 19	.1
San Antonio	Near Elmendorf, Tex.	June 11	<.1
Colorado	Near Yuma, Ariz.	June 18	<.1

devious pathways does it find its way again to sinks of a temporary or more permanent nature in deep soil, river and lake beds, and the ocean floor.

A very much simplified indication of the ramifications introduced by man is given in Fig. 2. One segment of the ecological complex that has received considerable attention in recent months is that created by the escape of metallic mercury from certain industrial plants such as those producing caustic soda and chlorine gas from common salt (Fig. 3). Tentative estimates of the relative quantities involved are shown in Fig. 4.

Unfortunately for a proper appreciation of these cycles and their significance for human health, insufficient information is available on the form -- metal, inorganic salts, organic compounds -- in which mercury is to be found at various parts of the cycle, or on the rate of conversion from one form to another during its passage. As will be seen later, the toxicity of mercurials for man varies greatly with

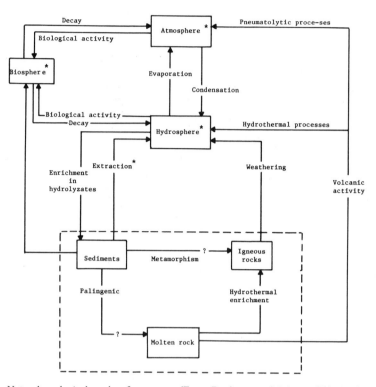

Fig. 1. Natural ecological cycle of mercury. (From Rankama and Sahama (7)). *Indicates points at which human perturbations may occur.

the form in which it is taken into the body, as well as with the route by which it enters. Failure to recognize the conversion of relatively innocuous metallic mercury into toxic methyl mercury in the sludge of lake and river bottoms permitted contamination of lakes from industrial sources to go uncontrolled for some time.

DISTRIBUTION IN ENVIRONMENT

Natural

LITHOSPHERE -- Reported concentrations in the "normal" lithosphere range from 3 ppb to 10 ppm, with most less than .25 ppm. (3,4,8-11)

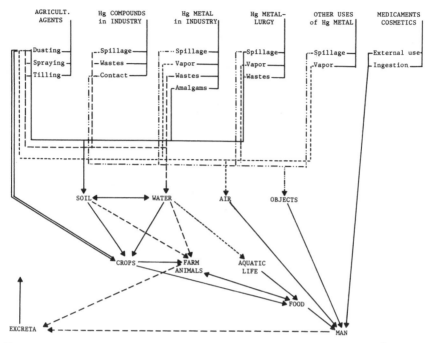

Fig. 2. Man-made impositions on the ecological cycle of mercury. (Supplied by D. H. K. Lee).

HYDROSPHERE -- Mercury's ubiquity probably applies to the hydrosphere as well as to the lithosphere. Presence of the element in seawater was demonstrated in 1799 (2) and it was found in a number of French mineral waters during the latter part of the 19th and early part of the 20th centuries (12-14). Prior to the 1960's the only important studies on mercury in the hydrosphere were those of Stock (3) as summarized in Table 1.

Extensive studies of mercury in the hydrosphere have been carried out by Soviet scientists, following the development of a rapid, sensitive, analytical method (4). Their investigations, published in 1962, covered some of the major rivers in Eastern Europe as well as the Black Sea, Sea of Azov, various parts of the Mediterranean Sea, and the Atlantic and Indian Oceans. Values for mercury ranged from 0.9×10^{-7} to 2.8×10^{-6} g/li (0.09 to 2.8 ppb) with little difference between the rivers and the seas.

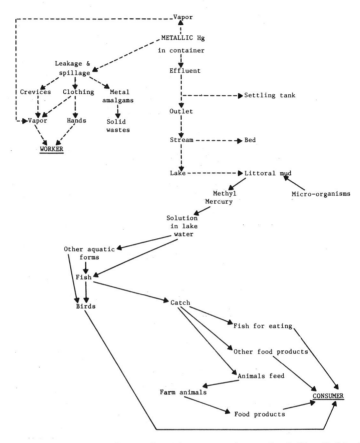

Fig 3. Flow paths of mercury from industrial sources to human food. (Supplied by D. H. K. Lee).

In the United States of America practically no attention was paid to mercury in the hydrosphere prior to 1970; earlier manuals and surveys dealing with water quality standards scarcely mention this element (15-17). As a result of this lack of interest it is hardly possible to know what the "natural" mercury content of water might have been. When studies were made in the spring and early summer of 1970, industrial and agricultural uses of mercurials had been common practice for at least two decades, so that many streams, rivers, and lakes may well have received some man-made infusions of mercury. Results of analyses performed by the U.S.

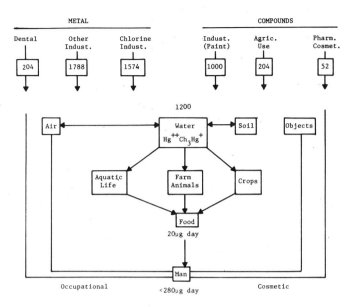

Fig. 4. Estimated rates of man-made imposition on the ecological cycle of mercury in the USA. Sources – thousands of pounds per year; incidence on man-micrograms per day.

Geological Survey (Table 2) suggest that the rivers sampled could not have received any large doses of the metal or its compounds, since most were found to contain less than 3 ppb. It was noted in the report, however, that while "... the natural concentration of mercury in most rivers and streams of the United States is 0.1 microgram per liter or less ... it may be several thousand times this concentration in some natural waters."

ATMOSPHERE -- As meagre as is the information on the "natural" mercury content of water, there is even less known about mercury in the atmosphere. On purely theoretical grounds it is reasonable to assume that mercury is present everywhere in the earth's atmosphere as a result of its relatively high vapor pressure at ordinary temperatures and the operation of the mercury cycle (Fig. 1). The metal may be present in the form of its vapor or as suspended particles (aerosols) of its compounds. All forms of mercury are capable of being converted to aerosols and thus may remain suspended in the atmosphere for indefinite periods of time.

Concentrations of mercury in the atmosphere under "natural" conditions are very low and consequently require the most sensitive methods for analysis, measurements being made in nanograms (10^{-9} grams). This being so, it is not surprising that studies on naturally-occurring atmospheric mercury have been limited in number and extent.

Late in 1960, S. H. Williston perfected an instrument which was capable of measuring mercury vapor in air in concentrations of 1-2 parts per trillion. Subsequent modifications have resulted in a model which is mobile and fully automated (18). Numerous observations made by Williston in the neighborhood of Palo Alto, California, have shown readings mainly in the range of 1 to 10 ng/m^3 of air, with occasional peaks up to 50 ng/m^3. These measurements do not include mercury in particulate form, that is, as dust or aerosol. Air samples taken at an elevation of 10,000 feet over the Pacific Ocean twenty miles off the California shore showed a mercury vapor content of 0.6-0.7 ng/m^3. This may be said to represent the "natural" background level of mercury vapor in the atmosphere, at least in one location. Comparable data for mercury in particulate form are not available. A limited study by Goldwater in New York in 1962 showed values of 0.001-0.041 mg/m^3 (1-40 ng/m^3).

BIOSPHERE -- Prior to the second half of the present century, there were very few studies of the "normal" amounts of mercury in plants and animals. General statements were made that plants and animals are able to concentrate mercury, in the case of marine algae up to more than 100 times that of sea water. Analyses by Stock and others (10,19-21) showed that many plants contain mercury in proportions that vary with the part. Gibbs (22) stated that mercury was found in many animal tissues without giving quantitative details.

Questions may properly be raised as to whether or not foods purchased in the open market and subjected to analysis do in fact represent "normal" conditions. Many foods undergo various types of processing before being offered for sale and there is no practicable way of determining possible removal or addition of mercury along the way. Furthermore, it would be virtually impossible to know how much mercury may have entered the foods as a result of practices in agriculture and animal husbandry. Comparisons of analyses of foods performed in the 1930's with more recent results are of some value, but must be made with caution in view of the employment of different analytical techniques.

Prior to 1970 several major studies had been conducted for the purpose of determining the mercury content of foods, the earliest in 1934 (3) and more recently in 1964 (23), with others in between (5,19,22). The results of these investigations are summarized in Table 3. Where comparisons are possible it would seem that there has been no significant change, but as mentioned above, the comparisons may not be entirely valid. In all of these studies, total mercury was measured with no attempt to determine its form.

Man-made

To the first simple forms of ecological redistribution resulting from the mining of cinnabar and subsequent extraction of the metal, countless others, both simple and complex, have been added, all offering some degree of danger to environmental integrity. The major distortions result from the discharge of wastes into air, water, or soil.

SOIL -- Few studies have been reported of concentrations resulting from contamination of soil with agricultural or industrial agents or wastes. Under some circumstances these could be high, and might eventually feed into and increase those of the hydrosphere. On the other hand, humic and fulvic acids in soils have strong mercury binding properties, which retard the leaching of mercury compounds. Agricultural and related pesticidal uses of mercury compounds constitute a significant man-made redistribution and consequently are important in affecting ecological balance. Many forms of inorganic and organic mercurials are used as pesticides (26). Corrosive sublimate ($HgCl_2$) is used as a seed disinfectant and to control many diseases of tubers, corms, and bulbs, including potatoes. Both mercuric and mercurous chloride are used to protect a number of vegetable crops, one advantage being their low cost. They are also used for the prevention and treatment of fungus diseases of turf grasses, especially on golf greens, but have been largely supplanted in these uses by phenyl mercurials.

Agricultural uses of organic mercury compounds were introduced into Europe around 1915 and into the United States in the early 1920's. The two most important applications are in treating seed to prevent fungal disease prior to germination and for the prevention of fungal diseases in growing plants, fruits, and vegetables. A variety of alkyl, aryl (phenyl), and alkoxy-alkyl compounds are in use.

WATER -- Man-made alterations in patterns of distribution of mercury in the hydrosphere have been of increasing concern in several parts of the world, at first in Japan and Sweden and later in the United States, Canada, and elsewhere. In Japan a mysterious illness, subsequently named Minamata disease, first made its appearance in the latter part of 1953, the highest incidence being among fishermen and their families (27-29). Later, when it was recognized that sea birds and household cats were being affected, attention was focused on fish and shellfish as etiological factors. This in turn led to a study of the water in Minamata Bay and to the identification of mercury in a factory effluent as the cause of the disease. At first it was thought that there had simply been a biochemical transformation of the inorganic into an alkyl (methyl) mercurial, either in fish or in the mud at the bottom of the bay. A change of this type has been shown to be possible (30-32). Ultimately, however, methyl mercury compounds were found along with the inorganic forms in the factory effluent (33). A second episode, similar but milder, occurred in the Agano River delta near Niigata in 1965 (34).

Late in 1967, the Swedish Medical Board found it necessary to ban the sale of fish from about 40 Swedish rivers and lakes, due to the finding of high concentrations of methyl mercury in fish caught in these waters. A general survey of surface waters revealed widespread pollution with organo-mercurials, but in most instances the levels were not sufficiently high to call for restrictive action.

The source of the mercury is believed to have been paper pulp mills which use mercurial compounds as slimicides. Serious poisoning of wild life, particularly game birds, had been noted in Sweden as early as 1960, the source probably being mercury-treated seed rather than polluted water, but recognition of the existence of a serious problem led to extensive research, which is still being continued. Much of what is now known about mercury in the environment is due to the work of Japanese and Swedish investigators (30,33,35).

A warning about mercury pollution of American rivers was sounded by the Sports Fishing Institute in August 1969 (36). This appears to have made little impact but when, in March 1970, a Norwegian scientist working in Canada announced the discovery of high concentrations of mercury in fish taken from Lake St. Clair, the reaction was immediate and dramatic. Widespread restrictions on fishing and the sale of fish have been instituted in many parts of the United States and Canada and drastic action has been taken by governmental agencies in both countries to abate the discharge of mercury-containing wastes into lakes and streams. Later studies revealed significant levels of mercury in tuna and swordfish, with the result that limitations were imposed on the sale of these species, particularly the latter.

AIR -- Up to the present time the discharge of mercury vapor or of mercury-containing dust into the atmosphere has been primarily a problem of industrial hygiene, and a very well known problem, too. One notorious exception to this, involving the general environment, is the situation described by Bernardino Ramazzini (1633-1714) in 1700 in which a citizen of the town of Finale, near Modena, sought an injunction against the operator of a factory making mercuric chloride, claiming that the fumes from the factory were causing an increased death rate in the town (37). All in all, mercury contamination of the general atmosphere has neither demanded nor received any major attention.

FOOD CHAIN -- Studies of the mercury content of foods between 1934 and 1964 (3,5,19,22,23) for the most part suggest that there was no significant change over this period (Table 3), but comparisons

TABLE 3 -- COMPARISON OF MERCURY CONCENTRATIONS IN FOODS

Concentrations (ppb) wet weight as reported by

Foodstuff	Stock (3) 1934	Stock (19) 1938	Gibbs (25) 1940	Columbia (24) 1964	Fujimura (5)* 1964
Meats	1.0- 67	5 - 20	.9- 43.6	1.2-150	310 -360
Fish	24 -180	25 -180	1.6- 13.6	0 - 59	35 -540
Vegetables (fresh)	2.0- 44	5 - 35	0	0 - 20	30 - 62
Vegetables (can)	.	.	5.0- 24.6	1.8- 7.2	.
Milk (fresh)	.6- 4.0	.6- 4.0	3.3- 7.1	8.1	3.2- 7.0
Butter	2	70 -280	.	141	.
Cheese	9 - 10	.	.	82	.
Grains	20 - 36	25 - 35	1.6- 6.0	2.3- 25.0	12 - 48
Fruits (fresh)	4.0- 12	5.0- 35	.	4.0- 31.0	18
Egg white	.	.	.	11	81 -125
Egg yolk	.	.	.	6.2	330 -670
Egg (whole)	2.2	2.0	0	.	.
Beer	.1- 1.4	1.0- 15	.	4.4	.

* The relatively high values reported by Fujimura may reflect the previous wide-spread agricultural use of mercurial pesticides in Japan.

are rendered difficult by variation in analytical techniques. Analyses made in 1970-1971, in response to alarm about the extent of mercury contamination, have brought to light considerably higher values in certain foods or food species (46). Newer analytical techniques should make values in the neighborhood of tolerance limits (0.5 ppm) more reliable than heretofore.

Levels of inorganic mercury in food and drinking water should not pose a threat at this time to the public's health (see later section on dose-response relationships). The presence of alkylmercury compounds in the human diet is, however, a matter for concern.

Mass poisonings by these compounds have caused many human fatalities and cases of permanent, incapacitating damage to the nervous system. Seed grain, treated with the organo-mercurial fungicide, ethylmercury p-toluene sulphon anilide, was used accidently in the preparation of bread and produced many cases of serious and fatal poisoning in Iran in the 1950's (91). Consumption of meat from livestock fed on treated grain has also caused a few serious cases of alkylmercury poisoning.

In addition to the agricultural uses of alkylmercury compounds, the manufacture of vinyl chloride from acetylene, using mercury salts as catalyst, has led to cases of human poisoning (as at Minamata Bay). Some 111 cases of poisoning were reported of which some were fatal. Fetal damage may have occurred in some cases even when the mothers were apparently healthy (38).

Mercury, in its elemental form and as inorganic salt, is a constituent of the earth's crust, and has been detected in all foodstuffs tested to date. However, the discovery of methylmercury compounds in various foodstuffs has given rise to extensive studies on the pathways by which this extremely dangerous compound of mercury enters food chains leading ultimately to man (39). Studies by Swedish scientists indicate that the levels of mercury in feathers of the goshawk and certain other birds rose dramatically after the introduction of alkylmercury compounds as seed dressing agents (40). The treated seed was consumed by small rodents which, in turn, were an important source of food for the birds. Surprisingly, it was subsequently discovered that mercury levels were also rising in feathers of birds not living or feeding in areas where alkylmercury compounds were used. Clearly, some widespread release of mercury must be responsible for these effects. Johnels and Westermark have suggested that levels in feathers of fish-eating birds probably coincide with increasing industrialization going back to the early nineteenth century (40).

The problem is made the more serious by the finding that mercury in fish is in the form of methylmercury compounds despite the fact that most mercury released into rivers, lakes, and oceans, is in the form of the inorganic salt or as the metallic element (39).

Jernelöv, (41) Wood et al., (42) have shown that microorganisms isolated from the sedimentary beds of rivers, canals, lakes, and even from aquaria are capable of methylating inorganic mercury. Mercury is thus converted into a highly diffusible form capable of leaving the sediment entering the water phase and thereby into a variety of organisms, including fish. The hazard to man is increased by the fact that most of the methylmercury found is in the edible portion of fish and the degree of accumulation in human brain is much greater than in most other species (40,43,44).

Westoo's observations (39) indicate that mercury in fish is in the form of monomethyl compounds but Wood et al.(42), claim that the dimethylmercury may be formed by methanogenic bacteria and that the reason that these compounds of mercury have not been detected is that the usual extraction procedures convert dimethyl to monomethyl compounds.

The quantitative significance of the biological methylation reactions described by Jernelöv, Wood, and others remains to be established. It offers the only explanation for the appearance of methylmercury compounds in fish caught in waters in areas where only metallic and inorganic forms of this element are known to be discharged. Large quantities of inorganic mercury have been released into many fresh water areas on the North American continent (45). Rising levels of mercury have been reported in fish caught in the Great Lakes and in smaller lakes in the North East (46). Unfortunately, there are no published reports identifying the chemical form of mercury in the fish specimens. According to the Scandinavian experience, we would expect all the mercury to be in the form of methylmercury compounds. On the other hand, studies on fish caught in Japanese waters indicate that methylmercury compounds account only for part of the total mercury in fish (10%-80%) (47). Clearly, there is an urgent need to identify the chemical form of mercury in fish caught in this contintent and also to check if industries using similar catalytical processes as those described in Japan are actively releasing methylmercury compounds.

The possibility has been raised that microorganisms present in the human gut may methylate inorganic mercury (116). However, this is a remote possibility in view of the fact that occupational exposure to inorganic or metallic mercury over periods of twenty years or more produce symptoms clearly distinct from those associated with exposure to methylmercury compounds (48).

Despite its apparent absence in mammalian tissues and the need for more studies on its quantitative significance, the discovery of biological methylation of inorganic mercury remains the most important single clue to understanding the movement of inorganic mercury into the food chain leading to ingestion by humans of methylmercury. The discovery serves as an important and timely warning that, if we are to assess the hazard to man resulting from the release of mercury and other metals into our environment, we must know more about the pathways into food.

It is customary to express the allowable daily intake of metals in food (the so-called ADI) in terms of the total amount of the element. No attempt is made to distinguish between the different chemical compounds of the metal in food. The experience in Japan and Sweden of food contamination with mercury clearly indicates that this approach is too crude. There are at least two reasons for this. First, the toxic effects of mercury differ with the chemical form of mercury as already discussed. Second, the site of absorption from the GI tract is greatly influenced by the chemical form of mercury. Studies in man and unpublished observations in experimental animals indicate that more than 90% of methylmercury is absorbed from food (44,49,50). In contrast, absorption of inorganic mercury salts is 50% or less depending on species and diet (50,51).

TOXICOLOGY

PHARMACOLOGICAL ASPECTS

Site of Action

Since its introduction as an antiseptic in 1881, the action of mercury on cells and cellular components has been widely studied. Voegtlin was the first to point to the importance of heavy metal interaction with -SH groups of proteins (52). Mercury has been shown to be present in urine complexed to the -SH group of cysteine (53) and to bind to the thiol group of a serum albumin (54).

A new general approach to the mechanism of action of heavy metals, pioneered by Rothstein (55), is based on the premise that the cell membrane is the first point of attack by heavy metals. In studies with uranium, copper, molybdenum, and finally mercury, he was able to show that heavy metals bind to the cell membrane and that the first detectable changes in cellular function are due to changes in the membrane -- either in inhibition of active transport processes or an increase in passive permeability. This membrane concept and the relevant studies up to 1960 have been the subject of a general review (56). The work with mercury has been reported only recently and indicates that this metal has a dramatic effect on membranes of a variety of mammalian cells (57-60).

It markedly increases the permeability to potassium and blocks the transport of sugars. These effects arise from a combination of mercury with -SH group in or on the cell membrane. Once inside the cell, the metal may be sequestered in an "inactive" combination, or it may react with enzymes or other sensitive "receptors" to elicit toxic effects. Weiner and co-workers have shown that only a fraction of the mercurial diuretic in kidney tissue is responsible for the diuretic effects (61).

Mercury is known to react with a large number of enzymes in vitro. Recent studies in sub-cellular distribution of mercury after administration to rats indicate that the metal is found in all parts of the cell (62). As a result there are many possibilities for enzyme damage. The difficulty of tracking down specific biochemical lesions that might be elicited by mercury in the whole animal is well illustrated by studies on the mechanism of action of mercurial diuretics. A variety of enzymes in renal tissues have been found to be inhibited by therapeutic or toxic doses of mercury chloride as the diuretic. However, in no case has it been possible to demonstrate any correlation between enzyme inhibition and changes in transport processes (63).

It should be noted that no specific biochemical test is available as a diagnostic aid in mercury poisoning. Both 3-aminolevulinic acid dehydratase activity of red blood cells, and cholinesterase activity of serum were decreased in workers exposed to mercury vapor and showing urinary mercury levels above 200 μg Hg/li (64), but there was no clear correlation with the degree of exposure.

Biotransformation

The rate of biotransformation of mercury and its compounds plays an important role in the absorption, distribution, organ deposition, and excretion. Two kinds of reactions have been studied; the interconversions of inorganic mercury between the elemental, and the mono and divalent states, and the cleavage of the carbon-mercury bond in organomercurial compounds. There is only one published account of the metabolic transformation of the organic moiety (65).

The vapor is oxidized in the body to divalent mercury. This oxidation has been shown to proceed in vitro on contact of the vapor with heparinized human blood. The oxidized mercury binds to serum albumin and to hemoglobin (66). The similarity of the general pattern of distribution and excretion of mercury inhaled as vapor or ingested as mercuric salts (67) suggests that vapor is rapidly oxidized in vivo to the divalent form.

However, brain levels are known to be higher after exposure to vapor. This is probably due to the fact that there will always be small amounts of dissolved vapor in the blood stream during the exposure period. This has now been demonstrated in experimental animals exposed to radioactive vapor (68). Due to its high lipid solubility and lack of charge, the dissolved vapor readily crosses the blood brain barrier. The vapor must be rapidly oxidized in the brain since mercury, once deposited in this organ, is slowly released.

Distribution and Excretion

The pattern of deposition of mercury is an important consideration in the toxicology of this metal. Both organic and inorganic mercury compounds, if present at a sufficient concentration, will poison any cell with which they come in contact. Irrespective of the route of administration, mercury is distributed to all organs of the body. The mercury in plasma is protein bound (69) and the mercury present in red blood cells is bound to the cysteine residues of hemoglobin (70,71). Mercury levels in kidney, expressed per gram wet weight, are nearly always higher than in any other organ (72,73).

The preferential renal deposition is probably the most important single process determining the overall pattern of distribution in the body tissues.

An efficient renal uptake helps to maintain the low concentration of mercury in plasma and in non-renal tissues. After injection of inorganic mercury salts and mercurial diuretics, mercury is rapidly removed from plasma and accumulated to high concentrations in the renal cortex; the mercury levels in other tissues are low. In contrast, the alkyl mercurials do not accumulate as well in kidney tissues but concentrations in other organs are relatively high (74). Any interference with the renal uptake process by complexing agents such as BAL or by metabolic inhibitors (2,4 dinitrophenol and sodium maleate) invariably leads to higher mercury levels in other tissues (75,76).

Despite the central role of renal deposition, very little is known about the mechanisms involved. Urinary excretion averaged over a group of workmen is generally accepted as a reliable index of exposure to inorganic mercury (77), but on an individual basis urinary excretion shows wide fluctuation from day to day.

Factors leading to variability in excretion will not be understood until more is known about the physiological processes involved in the renal handling of mercury. Probably no single mechanism is responsible for excretion. Literature reports based on animal data claim that mercury appears in urine by glomerular filtration (78), by secretion or diffusion across the tubular walls (79,80) or, in the case of toxic doses, by exfoliation of the tubular cells (81). The relative importance of these processes may depend upon such factors as the dose, the chemical form of mercury to which the animal is exposed, and the time after exposure.

Very little is known about the mechanisms of fecal excretion of heavy metals. This route of excretion is especially important in the case of exposure to methylmercury compounds. Recent studies by Norseth and Clarkson (82) have shown that 80% of the total excretion of mercury from rats injected with methylmercury chloride is by the fecal pathway.

CLINICAL ASPECTS

Health Effects

Metallic mercury in its liquid form is of little or no consequence as a threat to health. There is substantial evidence that it can be swallowed in amounts up to a pound or more with little adverse effect, thus negating the popular notion that metallic mercury as such is a deadly oral poison.

Inhalation of metallic mercury vapors can cause acute or chronic effects, which were recognized by Dioscorides as early as in the first century A.D. A number of cases of acute injury due to inhalation of mercury vapor have been reported, usually as a result of a household accident in which mercury has been heated (83-90). The principal effect is on the lungs in the form of an acute chemical pneumonitis leading to pulmonary edema and necrosis of pulmonary tissue. Symptoms resembling those of metal fume fever (chills, fever, cough, tight feeling in the chest) are common; some cases have ended fatally.

Chronic poisoning due to the inhalation of mercury vapor is far more common than the acute type, most cases being of occupational origin. The major manifestations are seen in the form of the classical triad of tremor, erethism, and gingivitis, although a host of non-specific subjective complaints have been described.

It is important to realize that metallic mercury will vaporize at ordinary room temperatures, the rate of vaporization increasing as the temperature rises.

Best known of the soluble salts of mercury is its bichloride ($HgCl_2$), also known as mercuric chloride, corrosive sublimate or simply sublimate. The toxic properties of mercuric chloride have been known for at least a thousand years and its use for homicidal and suicidal purposes is common knowledge. Taken by mouth this compound causes corrosion of the intestinal tract, leading to bloody diarrhea. Its absorption results in injury to the kidneys with suppression of urine and ultimate uremic death. Chronic poisoning due to bichloride of mercury has

35

not been prominent even though this compound was for many years administered orally as one form of treatment for syphilis.

Mercurous chloride (HgCl, calomel) is less soluble than the mercuric form and therefore less dangerous. It has been and still is used medicinally, but its etiological role in causing acrodynia in infants has led to curtailment of some of its uses. Red oxide of mercury, used in some anti-fouling paints for ship bottoms, may contribute in a minor degree to water pollution, but it, as well as other inorganic mercurials, are of relatively little importance in environmental toxicology.

Of the several organic forms, the alkyl mercurials are causing the greatest concern at the present time. Human poisoning due to the ingestion of grain treated with mercurial fungicides has been reported from different parts of the world. Outbreaks involving hundreds of persons occurred in Iraq in 1956 and 1960 (91), with ethylmercury p-toluene sulphonanilide as the offending agent. The exact number of cases of poisoning is not known, but in the 1956 outbreak more than 100 cases, with 14 deaths, were admitted to Mosul Hospital and in the 1960 tragedy, 221 patients were treated in one hospital in Baghdad (92). A similar outbreak occurred in Guatemala in 1965, involving 45 cases with 20 deaths (93). In 1969 three members of a New Mexico family became seriously ill as a result of eating meat from hogs which had been fed on grain that had been treated with methylmercury dicyandiamide (94).

The clinical manifestations seen in the episodes mentioned above and in several dozen sporadic cases reported in the literature (95-101) lead to the conclusion that most, if not all, alkyl mercurials behave similarly when taken into the human body. The predominant feature is permanent organic injury to the brain resulting in weakness progressing to paralysis, loss of vision, and disturbed cerebral function. Severe cases become comatose and end fatally.

Congenital defects in the form of cerebral palsy (102) and mental retardation (103) have been associated with alkyl mercury poisoning, and in the laboratory it has been shown that chromosomal abnormalities may be produced (104,105).

Extensive, continuing studies on humans exposed to phenyl-mercurials were initiated at Columbia University in 1961 with reports on the findings being published from time to time (106-108). Most of the subjects under study were in fact handling mixtures of inorganic and phenyl compounds, but the predominant exposure was to the latter. Limited observations on humans whose contact was with unmixed phenylmercurials revealed responses similar to those with the mixed exposures (106-5). Noteworthy among the findings has been the almost complete absence of any abnormalities of clinical significance that could be attributed to the absorption of mercury, in spite of exposures greatly in excess of accepted Threshold Limit Values. Similar minimal responses have been observed in experimental animals (109).

Relatively few articles dealing with the toxicity of alkoxy alkyl mercurials have appeared in the literature and some of these are of little value. It would appear that it is not possible to generalize about the toxicology of alkoxy alkyl mercury compounds (65,110).

Dose-Response Relationships

Establishment of exact relationships between the dose of a mercurial taken into the body and the health effects to be expected has encountered several difficulties. Until recently emphasis has been placed on the amount of the element taken in, appearing in the tissues, or excreted in the urine. It is now realized, as was pointed out in a preceding section, that the nature of the compound in which the mercury is included has a marked effect upon absorption and metabolism of the element, and therefore upon its toxicity. Data which refer only to the amount of elemental mercury are not of much help in determining dose-response relationships. Methods of analysis, moreover, have not been very sensitive, so that the effects of small doses continued over long periods could not be followed with accuracy. And finally, the factors that determine the biotransformation of mercurials, their passage through barriers within the body, and the ultimate action upon cellular mechanisms are only now beginning to be understood. The evidence upon which concepts of the dose-response relationships can be constructed is, therefore, indirect, not always replete with detail, and sometimes contradictory. A lot of it is negative -- that a certain dose did not produce demonstrable effects, leaving open the question whether the right evidence of effects was sought.

The amount of a particular compound present in the body and thus able to exert toxic effects at a particular time is the result of a balance between intake and excretion. When the same amount is taken in each day, the body content rises progressively to a plateau at which the excretion equals the intake. Fig. 5 indicates this for mice. The time to reach steady state body levels is determined by the half-time of excretion. Taking the half-time of excretion in man as 70 days, steady state levels in the human population will be reached in approximately one year. Once attained, the steady state level of mercury is proportional to the daily intake. Studies on blood levels in humans exposed to methylmercury support this conclusion. The steady state blood level in mice was proportional to the concentration of mercury in the food providing exposure had been for one year or more.

This relationship between food and blood levels has been used by Berglund and Berlin (44) to estimate allowable daily intake of methylmercury compounds. They concluded that no-effect intakes for man should be in the range of 0.6 to 1.0 mg/day with the corresponding allowable daily intakes one-tenth of this. Their calculations are based ultimately on neurological effects in adult humans recorded during the epidemics in Japan.

They do not take into account, however, fetal damage by methylmercury compounds. The epidemiological data for Japan indicate that apparently healthy females may give birth to offspring with brain damage (38,111). Studies in Sweden reviewed by Lofroth indicate that human fetal blood levels are higher than levels in maternal blood (38), and animal experiments indicate that

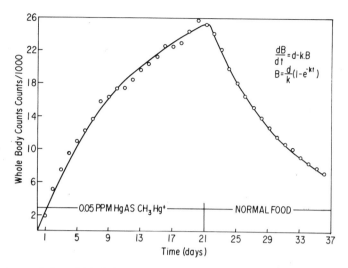

Fig. 5. Accumulation of labelled mercury in the mouse given a constant daily dose in food, and rate of elimination when dose is discontinued.

methylmercury salts damage the fetus. It is not known what levels in fact are no-effect levels with respect to fetal damage.

The estimation of hazard to man is made the more difficult by our lack of knowledge of interactions of alkylmercury compounds with drugs. The cases reported in Japan were on females whose drug intake during pregnancy was not known. In the Western Hemisphere it has been reported that pregnant females may receive up to a dozen different drugs during pregnancy (112). The possible interactions between methylmercury compounds and drugs commonly used during pregnancy is a matter deserving urgent investigation.

Species variation in excretion half-times (Table 4) has an important bearing on the accumulation of mercury in food chains. For any given dietary intake of mercury, the body levels are directly proportional to the biological half-times. Many aquatic organs have long half-lives and accordingly accumulate methylmercury. For example, the seal has a biological half-time of approximately 500 days, and has been reported to contain high levels of mercury.

39

TABLE 4 -- BIOLOGICAL HALF-TIMES OF METHYLMERCURY
IN DIFFERENT SPECIES

Species	Single Dose	Half-Time in days	Ref.
Mouse	oral, i.p.	8	50
Rat	oral, i.v.	16	113, 114
Squirrel Monkey	oral	65	44
Seal	oral	500	115
Poultry	oral	25	116
Molluscs	i.m.	481-1000	117
Crab	i.v.	400	117
Pike	oral, i.m.	640- 780	118
Flounder	oral, i.m.	700-1200	118
Eel	oral, i.m.	910-1030	118

The mathematical considerations involved in the above experiment are as follows:

The points in the figure are the means of measurement on five mice (CBA strain). On returning the animals to normal food, the levels of mercury decline. The curve relating the amount in the body as a function of time may be represented by the equation (50):

$$B = (d/k) \{1 - \exp (-kt)\} ------------(1)$$

where B is the amount of mercury in the body, at any time t, d is the constant amount absorbed each day from the food and k is the rate constant for excretion of mercury defined as

$$\text{rate of excretion} = kB ----------------(2)$$

The continuous curve in Fig. 5 is drawn according to equation (1). The value of d is identical with the amount of mercury ingested indicating complete absorption of the methylmercury from food.

If the animals are maintained on radioactive food for a long period of time, the body burden of mercury becomes steady corresponding to the point where excretion is equal to absorption. The value of the steady state body burden, B_∞, according to equation (1) is:

$$B_\infty = d/k ---------------------------(3)$$

The half-time of excretion of mercury, $t_{1/2}$, is related to k according to the expression:

$$t_{1/2} = .693/k \text{ -----------------------(4)}$$

Thus we may rewrite equation (3) as:

$$B_\infty = .693d.t_{1/2} \text{ -------------------(5)}$$

The uptake and excretion of mercury compounds in humans follows a similar kinetic pattern. Differences between species appear to be restricted to difference in the value of the biological half-times -- see Table 4.

Little information is on hand from which to assess the potential hazard of inorganic mercury in food. No epidemiologic data is available. Industrial exposures indicate that air levels of the order of 0.1 mg/m^3 are at the no-effect threshold (48). This would correspond to an intake of 5 mg Hg/week or 0.7 mg/day. Mercury vapor is absorbed in the lung with an efficiency of 80% (Fiserova-Bergerova (119)) and, according to observations referred to previously, the GI tract absorbs from 2 to 50% of inorganic mercury depending upon diet and animal species. If the absorption from food is as high as 50%, the no-effect food intake would be approximately 1 mg/day. Despite the gross assumptions in the calculation, this figure is comfortably distant from the total intake in food of 25 µg/day. Stokinger and Woodward (120) have made similar calculations of safe levels in drinking water using threshold limit concentrations in air and have discussed the assumptions involved.

Diagnostic Criteria

The no-effect level for methylmercury compounds in blood is believed to be 60 µg/100 ml based on data from epidemics of human poisoning in Japan (121). However a recent Swedish analysis of the Japanese data suggest that the no-effect level may be as low as 20 µg/100ml (123).

Insofar as the limited observations permit, the evidence indicates that blood levels of methylmercury

41

are a good index of exposure providing people have been on a more or less steady intake for at least one year.

These no-effect levels, with the exception of methylmercury compounds, are based on studies of occupational exposure, but cannot be applied to the general population without additional safety factors. Occupationally derived threshold limit values and maximal allowable concentrations do not take into account the extremes of youth, age, and disease encountered in the general population (120).

In the case of inorganic salts of mercury, the relationship between blood levels and body burdens, or between blood levels and urinary excretion, is much more complicated than in the case of methylmercury compounds. Present evidence indicates that plasma concentrations of the metals should be considered rather than whole blood levels. Vostal's recent discussion of the relationship between blood levels and urinary excretion of different compounds of mercury well illustrates this point (122). A graph of daily urinary excretion versus whole blood concentrations gave approximate linear relationships in the case of people occupationally exposed to mercury compounds. However, the slopes of the lines were widely different between people exposed to alkyl mercury compounds and those exposed to inorganic salts or to the vapor. By making use of the known distribution ratios of the metal between plasma and red blood cells for the different chemical compounds of mercury, Vostal was able to replot the data in terms of plasma concentrations versus urinary excretion. In this case, all the data fell in the same straight line over the entire range of urinary excretion rates and plasma concentrations.

The need to measure plasma concentrations, or to distinguish between different chemical compounds of metals in blood, offers a difficult challenge to the analyst. Most of the present analytical techniques possess neither the necessary sensitivity nor the selectivity. For example, in a recent study of mercury levels in the general population using one of the most sensitive techniques available, over 75% of the mercury concentrations in blood were below the limit of detection (77).

REFERENCES

1. Aller, L. H. (1961). The Abundance of the Elements. Interscience Publishers, New York.

2. Proust, J. L. (1799). On the existence of mercury in the waters of the ocean. J. Phys. 49: 153.

3. Stock, A. and Cucuel, F. (1934). Die Verbreitung des Quecksilbers. Naturwissenschaften 22/24: 390.

4. Aydin'yan, N. K. (1962). The content of mercury in certain natural waters. Transactions of the Institute of the Geology of Ore Deposits, Petrography, Mineralogy and Geochemistry, Moscow, No. 70: 9.

5. Fujimura, Y. (1964). Studies on the toxicity of mercury (Hg Series No. 7). 2nd report. On the present status of mercury contamination in environment and foodstuffs. Jap. J. Hyg. 18: 10.

6. U. S. Geological Survey. (1970). Water Resources Review: 7.

7. Rankama, K. and Sahama, T. G. (1950). Geochemistry. Chicago University Press, Chicago.

8. Preuss, A. (1940). Spectrographic methods. Zeitschr. angew. Mineralogie 3: 8.

9. Saukov, A. A. (1946). Geochemistry of Mercury. Trans. Instit. Geol. Sci. No. 78.

10. Warren, H. V., Delavault, R. E. and Barasko, J. (1966). Some observations on geochemistry of mercury as applied to prospecting. Economic Geology 61: 1010.

11. Andersson, A. (1967). Mercury in Swedish soils. Oikos, Sup. 9: 13.

12. Garrigou, F. (1877). Sur la presence du mercure dans la source du Rocher. Compt. Rend. 84: 963.

13. Willm, E. (1879). Sur la presence du mercure dans les eaux de Saint-Nectaire. Compt. Rend. 88: 1032.

14. Bardet, J. (1913). Etude Spectrographique des eaux minerales francaises. Compt. Rend. 157: 224.

15. McKee, J. E. and Wolf, H. W. (1963). Water Quality Criteria. (2nd ed.) Resources Agency of California, State Water Quality Control Board, Pub. 3-A.

16. Water Quality Criteria. (1968). Federal Water Pollution Control Admin.

17. Kopp, J. F. and Kroner, R. C. (1968). Trace metals in waters of the United States. U. S. Dept. of the Interior Federal Water Pollution Control Administration, Division of Pollution Surveillance, Cincinnati.

18. Williston, S. H. (1968). Mercury in the atmosphere. J. Geophys. Res. 73: 7051.

19. Stock, A. (1938). Die microanalytische Bestimmung des Quecksilbers und ihre Anwendung auf hygienische und medizinische Fragen. Svensk Kem Tidskr 50: 242.

20. Zimmerman, P. W. and Crocker, W. (1934). Plant injury caused by vapors of mercury. Contribs. Boyce Thompson Inst. 6: 167.

21. Dimond, A. E. and Stoddard, E. M. (1955). Toxicity to greenhouse roses from paints containing mercury fungicides. Conn. Agric. Exp. Sta. Bull. No. 595.

22. Gibbs et al. (1941). Absorption of externally applied mercury. Arch. Dermatol. and Syphil. 44: 862.

23. Cerkez, F. and Goldwater, L. J. (1964). Unpublished observations.

24. Cerkez, F. (1964). The evaluation of mercury exposure in human environment with emphasis on content in food. Columbia University School of Public Health (Dissertation).

25. Gibbs, O. S., Shank, R. and Pond, H. (1940). Report of studies on ammoniated mercury ointments and red oxide of mercury ointments. The External Products Research Institute, Memphis, Tennessee. (Processed)

26. Kirk-Othmer Encyclopedia of Chemical Technology. (1967). John Wiley and Sons, New York (2nd ed.) 13.

27. Kurland, L. T., Faro, S. N. and Siedler, H. (1960). Minamata Disease. World Neurology 1: 371.

28. Takeuchi, T. A. (1961). A pathological study of minamata disease in Japan. Symposium on Geographic Neurology in Conjunction with the 7th International Congress of Neurology, Rome, 1.

29. Tokuomi, H. et al. (1961). Minamata disease -- An unusual neurological disorder occurring in Minamata, Japan. Kumamoto Med. J. 14: 47.

30. Kondo, T. (1964). Studies on the origin of the causative agent of minamata disease. 4. Synthesis of methyl (methylthio) mercury. J. Pharmaceut. Soc. Japan 84: 137.

31. Wood, J. M., Kennedy, F. S. and Rosen, C. G. (1968). Synthesis of methylmercury compounds by extracts of a methanogenic bacterium. Nature 220: 173.

32. Jernelöv A. (1969). A conversion of mercury compounds. In: M. W. Miller and G. G. Berg (eds.) Chemical Fallout. Charles C. Thomas, Springfield, Illinois. 68.

33. Irukayama, K. et al. (1962). Studies on the origin of the causative agent of minamata disease. 3. Industrial wastes containing

mercury compounds from Minamata factory. Kumamoto Med. J. 15: 57.

34. Tsuchiya, K. (1969). Personal Communication.

35. Various Authors. (1967). The mercury problem. Oikos, Sup. 9.

36. Water Newsletter. (1969). Water Information Center, Inc. 11: 16.

37. Ramazzini, B. (1964). De Morbis Artificum. Trans. Wilmer Cave Wright, Hafner Publishing Company, New York.

38. Lofroth, G. (1969). A review of health hazards and side effects associated with the emission of mercury compounds into natural systems. Ecological Research Committee Bulletin No. 4. Swedish Natural Science Research Council. Sueauagen 166 8th, 11236, Stockholm.

39. Westöö, G. (1969). Methyl mercury compounds in animal foods. In: M. W. Miller and G. G. Berg (eds.) Chemical Fallout. Charles C. Thomas, Springfield, Illinois. 75.

40. Johnels, A. G. and Westmark, T. Mercury contamination of the environment in Sweden. In: M. W. Miller and G. G. Berg (eds.) Chemical Fallout. Current research in persistent pesticides. Charles C. Thomas, Springfield, Illinois. 221.

41. Jernelöv, A. (1969). Conversion of mercury compounds. In: M. W. Miller and G. G. Berg (eds.) Chemical Fallout. Charles C. Thomas, Springfield, Illinois. 68.

42. Wood, J. M., Kennedy, F. S. and Rosen, C. G. (1968). Synthesis of methyl mercury compounds by extracts of a methanogenic bacterium. Nature 220: 173.

43. Eckman, L. et al. (1968). Metabolism and retention of methyl-203-mercury nitrate in man. Nordisk Medicin 79: 456.

44. Berglund, F. and Berlin, M. (1969). Risk of methyl mercury cumulation in man and mammals and the relation between body burden of methyl mercury and toxic effects. In: M. W. Miller and G. G. Berg (eds.) Chemical Fallout. Charles C. Thomas, Springfield, Illinois. 258.

45. Somers, E. and Smith, D. M. (1971). Source and occurrence of environmental contaminants. J. Food Cosmet. Toxicol. 9: 185.

46. Jervis, R. E. et al. (1969-70). Mercury residues in Canadian foods, fish and wildlife. Natural Health Grant Project No. 605-7-S10. Trace Mercury in Environmental Materials.

47. Ueda, K. and Aoki, H. (1970). Methyl and ethyl mercury in the fishes of fresh water in Japan. Jap. J. Hyg. 25: 1.

48. Karolinska Institute. (1969). Maximum allowable concentrations of mercury compounds. Arch. Env. Health 19: 891.

49. Aberg, B. et al. (1969). Metabolism of methyl mercury (203 Hg) compounds in man. Arch. Env. Health 19: 478.

50. Clarkson, T. W. (1971). Epidemiological and experimental aspects of lead and mercury contamination of food. J. Food and Cosmet. Toxicol. 9: 229.

51. Fitzhugh, O. G. et al. (1950). Chronic oral toxicities of mercury-phenyl and mercuric salts. Arch. Indust. Hyg. Occup. Med. 2: 433.

52. Voegtlin, C., Dyer, H. A. and Leonard, C. S. (1923). On the mechanism of the action of arsenic upon protoplasm. U. S. Public Health Service Rep. 38: 1882.

53. Weiner, I. M., Levy, R. I. and Mudge, G. H. (1962). Studies on mercurial diuresis; renal excretion, acid stability and structure activity relationship of organic mercurials. J. Pharmacol. Exp. Therap. 138: 96.

54. Hughes, W. L., Jr. (1950). Protein mercaptides. Cold Spring Harbor Symp. Quant. Biol. 14: 79.

55. Rothstein, A. (1959). The cell membrane as the site of action of heavy metals. Federation Proc. 18: 1026.

56. Passow, H. Rothstein, A. and Clarkson, T. W. (1961). The general pharmacology of heavy metals. Pharmacol. Rev. 13: 185.

57. Weed, R. I., Eber, J. and Rothstein, A. (1962). Interaction of mercury with human erythrocytes. J. Gen. Physiol. 45: 395.

58. Van Steveninck, J. R., Weed, R. I. and Rothstein, A. (1965). Localization of erythrocyte -- membrane sulfhydryl groups essential to glucose transplant. J. Gen. Physiol. 48: 617.

59. Sutherland, R., Rothstein, A. and Weed, R. I. (1967). Erythrocyte membrane sulfhydryl groups essential for normal cation permeability. J. Cell Physiol. 69: 185.

60. Clarkson, T. W. and Cross, A. C. (1961). Studies of the action of mercuric chloride on intestinal absorption. AEC Research and Development Report U. R. 588.

61. Weiner, I. M. et al. (1959). The effect of dimercaprol (BAL) on the renal excretion of mercurials. J. Pharmacol. Exp. Ther. 127: 325.

62. Norseth, T. (1968). The intracellular distribution of mercury in rat liver after a single injection of mercuric chloride. Biochem. Pharmacol. 17: 581.

63. Webb, J. L. (1966). Enzyme and Metabolic Inhibitors. 2, Academic Press, New York.

64. Wadu, O. et al. (1969). Response to a low mercury concentration of mercury vapor. Arch. Env. Health 19: 485.

65. Daniel, J. W., Gage, J. C. and LeFevre, P. A. (1971). Metabolism of methoxyethyl mercury salts. Biochem. J. 121: 411.

66. Clarkson, T. W. (1965). Mercury -- Toxicological aspects. Ann. Occup. Hyg. 8: 73.

67. Rothstein, A. and Hayes, A. (1964). The turnover of mercury in rats exposed repeatedly to inhalation of vapor. Health Physics 10: 1099.

68. Magos, L. (1967). Mercury-blood interaction and mercury uptake by the brain after vapor exposure. Env. Res. 1: 323.

69. Cember, H., Gallerher, P. and Faulkner, A. (1968). Distribution of mercury among blood group fractions and serum proteins. Am. Ind. Hyg. Assoc. J. 29: 233.

70. Weed, R., Eber, J. and Rothstein, A. (1962). Interaction of mercury with human erythrocytes. J. Gen. Physiol. 45: 395.

71. Takeda, Y. et al. (1968). Mercury compounds in the blood of rats treated with ethylmercuric chloride. Toxic. and Appl. Pharmacol. 13: 165.

72. Rothstein, A. and Hayes, A. L. (1964). The turnover of mercury in rats exposed repeatedly to inhalation of vapor. Health Physics 10: 1099.

73. Kessler, R. H., Lozano, R. and Pitts, R. F. (1957). Studies on structure, diuretic activity relationships of organic compounds of mercury. J. Clin. Invest. 36: 656.

74. Swensson, A. and Ulfarson, U. (1967). Toxicology of organic mercury compounds used as fungicides. Occup. Health Rev. 15: 5.

75. Clarkson, T. W. and Magos, L. (1967). The effect of sodium maleate on the renal deposition and excretion of mercury. Brit. J. Pharmacol. 31: 560.

76. Clarkson, T. W. and Magos, L. (1970). Effect of 2,4-dinitrophenol and other metabolic inhibitors on the renal deposition and excretion of mercury. Biochem. Pharmacol. 19: 3029.

77. Goldwater, L. J. (1964). Occupational exposure to mercury, The Harben Lectures. J. Roy. Inst. Public Health & Hyg. 27: 279.

78. Wokel, W., Stegner, H. E. and Janisch, W. (1961). Zum topochemischer Queck Silbernachweis in der Niere bei Experimental Sublimat Vergiftung. Virchows. Arch. Path. Anat. 334: 503.

79. Mambourg, A. M. and Raynard, C. (1965). Etude a l'aide d'isotopes radioactifs du mecanisme de l'excretion urinaire du mercure chez le lapin. Revue Etud Clin. Biol. 10: 414.

80. Vostal, J. (1966). Study of renal excretory mechanisms of heavy metals. Proc. 15th Int. Cong. Occup. Health, Vienna 3: 61.

81. Cember, H. (1962). The influence of the size of dose on the distribution and elimination of inorganic mercury $Hg(NO_3)_2$ in the rat. Ind. Hyg. J. 23: 304.

82. Norseth, T. and Clarkson, T. W. (1971). Intestinal transport of [203]Hg-labelled methyl mercury chloride. Role of biotransformation in rats. Arch. Env. Health 22: 568.

83. Campbell, J. S. (1948). Acute mercuric poisoning by inhalation of metallic vapour in an infant. Canad. Med. Assoc. J. 58: 72.

84. King, G. W. (1954). Acute pneumonitis due to accidental exposure to mercury vapor. Ariz. Med. 11: 335.

85. Matthes, F. T. et al. (1958). Acute poisoning associated with inhalation of mercury vapor. Pediatrics 22: 675.

86. Teng, C. T. and Brennan, J. C. (1959). Acute mercury vapor poisoning. Radiology 73: 354.

87. Haddad, J. K. and Stenberg, E. (1963). Bronchitis due to acute mercury inhalation. Amer. Rev. Resp. Dis. 88: 543.

88. Hallee, J. T. (1969). Diffuse lung disease caused by inhalation of mercury vapor. Amer. Rev. Resp. Dis. 99: 430.

89. Tennant, R., Johnston, J. H. and Wells, J. B. (1961). Acute bilateral pneumonitis associated with the inhalation of mercury vapor. Conn. Med. 25: 106.

90. Hopmann, A. (1927). Acute Quecksilberdampfvergiftungen. Zentrlblatt f. Gewerbehyg. 4: 422.

91. Jalili, M. A. and Abbasi, A. H. (1961). Poisoning by ethyl mercury toluene sulphonanilide. Brit. J. Indust. Med. 18: 303.

92. Damluji, S. (1962). Mercurial poisoning with the fungicide granosan. M. J. Facul. Med. Baghdad 4: 83. Abstr. in Bull. Hyg. 38: 155 (1963).

93. Anon. (1966). Mercury poisoning in Guatemala. Morbidity and Mortality Weekly Report 15: 34. (U. S. Dept. of HEW, PHS, CDC).

94. Storrs, B. et al. (1970). Organic mercury poisoning-- Alamogordo, New Mexico. Morbidity and Mortality Weekly Report 19: 25.

95. Koelsch, F. (1937). Gesundheitsschadigungen durch organische Quecksilberverbindungen. Archiv Gewerbepath. Gewerbehyg. 8: 113.

96. Swensson, A. (1952). Investigations on the toxicity of some organic mercury compounds which are used as seed disinfectants. Acta Med. Scand. 143: 365.

97. Bidstrup, P. L. (1964). Toxicity of Mercury and its Compounds. Elsevier Publishing Company, Amsterdam, London and New York.

98. Teleky, L. (1955). Gewerbliche Vergiftungen. Springer-Verlag, Berlin, 96.

99. Zangger, H. (1930). Erfahrungen über Quecksilbervergiftungen. Archiv Gewerbepath. Gewerbehyg. 1: 539.

100. Katsunuma, H. et al. (1963). Four cases of occupational organic mercury poisoning. Resp. Inst. Sci. Labor (Japan) No. 61: 33.

101. Koelsch, F. (1950). Vergiftungen durch organische Quecksilber-Verbindungen. Proc. Internat. Congress of Occup. Med. Milan, 103.

102. Matsumoto, H. G., Koya, G. and Takeuchi, T. (1965). Fetal minamata disease -- A neuropathological study of two cases of intrauterine intoxication by a methyl mercury compound. J. Neuropath. Neurol. 24: 563.

103. Englesson, J. and Herner, A. B. (1952). Alkyl mercury poisoning. Acta Paediat. 41: 289.

104. Ramel, C. (1967). Genetic effects of organic mercury compounds. Oikos, Sup. 9: 35.

105. Skerfving et al. (1970). Chromosome breakage in humans exposed to methyl mercury. Arch. Env. Health 21: 133.

106. Goldwater, L. J. et al. Absorption and excretion of mercury in man:

    1. Relationship of mercury in blood and urine, Arch. Env. Health 5: 537 (1962).

    2. Urinary mercury in relation to duration of exposure, Arch. Env. Health 6: 480 (1963).

3. Blood mercury in relation to duration of exposure, Arch. Env. Health 6: 634 (1963).

4. Tolerance to Mercury, Arch. Env. Health 7: 568 (1963).

5. Toxicity of phenylmercurials, Arch. Env. Health 9: 43 (1964).

6. Significance of mercury in urine, Arch. Env. Health 9: 454 (1964).

7. Significance of mercury in blood, Arch. Env. Health 9: 735 (1964).

8. Mercury exposure from house paint -- A controlled study, Arch. Env. Health 11: 582 (1965).

12. Relationship between urinary mercury and proteinuria, Arch. Env. Health 15: 155 (1967).

13. Effects of mercury exposure on urinary excretion of coproporphyrin and delta-aminolevulinic acid, Arch. Env. Health 15: 327 (1967).

14. Salivary excretion of mercury and its relationship to blood and urine mercury, Arch. Env. Health 17: 35 (1968).

107. Goldwater, L. J. (1964). Occupational exposure to mercury: The Harben Lectures. J. Roy. Inst. Public Health and Hyg. 27: 279.

108. Goldwater, L. G. (1966). Occupational exposure to mercury. Proc. 15 International Cong. on Occupational Health, Vienna. (Paper BIII-17).

109. Gage, J. C. and Swan, A. A. B. (1961). The toxicity of alkyl and aryl mercury salts. Biochem. Pharmacol. 8: 77.

110. Norseth, T. (1967). The intracellular distribution of mercury in rat liver after

methoxymethyl mercury intoxication. Biochem. Pharmacol. 16: 1645.

111. Takeuchi, T. (1970). Biological reactions and pathological changes of human beings and animals under the condition of organic mercury poisoning. Presented at the International Conference on Environmental Mercury Contamination in Ann Arbor.

112. Bleyer, W. et al. (1970). Studies on the detection of adverse drug reactions in the new born. J. Amer. Med. Assoc. 213: 2046.

113. Norseth, T. (1970). Studies of the biotransformation of methylmercury salts in the rat. Ph.D. thesis, Univ. of Rochester, New York.

114. Gage, J. C. (1964). Distribution and excretion of methyl and phenyl mercury salts. Brit. J. Ind. Med. 21: 197.

115. Tillander, M. and Miettinen, J. K. (1970). Excretion rate of methylmercury in the seal. FAO. Technical Conference on Marine Pollution. FIR: MP/70/ E-67 Rome, Italy.

116. Swensson, A. and Ulfarson, U. (1968). Distribution and excretion of various mercury compounds after single injections in poultry. Acta Pharmacol. and Toxicol. 26: 259.

117. Miettinen, J. K., Heyrand, M. and Keckes, S. (1970). Mercury as a hydrospheric pollutant. 1. Biological half-time of methyl mercury in four Mediterranean species: A fish, a crab and two molluscs. FAO Technical Conference. FIR: MP/70/ E-90. Rome, Italy.

118. Jarvenpaa, T., Tillander, M. and Miettinen, J. K. (1970). Methylmercury: Half-time of elimination in flounder, pike, and eel. FAO Technical Conference. FIR/70/ E-66. Rome, Italy.

119. Teisinger, J. and Fiserova-Bergerova, V. (1965). Pulmonary retention and excretion of

mercury vapors in man. Ind. Med. Surg. 34: 580.

120. Stokinger, H. E. and Woodward, R. L. (1958). Toxicologic methods for establishing drinking water standards. J. Amer. Water Works Assoc. 50: 515.

121. Berglund, F. and Berlin, M. (1969). Human risk evaluation for various populations in Sweden due to methyl mercury in fish. In: M. W. Miller and G. G. Berg (eds.) Chemical Fallout. Charles C. Thomas, Springfield, Illinois. 75.

122. Vostal, J. (1968). Renal excretory mechanisms of mercury compounds. Working paper for the Symposium on MAC Values, Stockholm. Arch. Env. Health 19: 891.

123. Berglund, F. et al. (1970). Methyl mercury in fish. A toxicologic-epidemiologic evaluation of risks. Nordisk Hygienisk Tidskrift Suppl. 4.

CHAPTER 3. LEAD

ROBERT A. GOYER, Department of Pathology,
University of North Carolina School of Medicine,
Chapel Hill, North Carolina

J. JULIAN CHISOLM, Department of Pediatrics,
Baltimore City Hospital, Baltimore, Maryland

Classical lead poisoning has been well known for centuries. Nevertheless, lead has continued to be one of man's most useful metals. Concern has grown in recent years over the questions of whether or not the increased use and dissemination of lead through the environment is resulting in undetected, subclinical, adverse effects on health.

ECOLOGY

SOURCES

Geological

Lead usually occurs either as sulfide (galena), oxide or carbonate ores (anglesite, cerussite). Compounds of iron, zinc, silver, copper, gold, cadmium, antimony, arsenic, bismuth, and other metals may be associated in various proportions. The average lead content of mined ores ranges from 3 to 8% (1).

Industrial

Mining, smelting, refining, secondary recovery, use of lead-containing products, and waste disposal result in dissemination of lead in the environment. Mine production of recoverable lead in the United States has exceeded 250,000 tons annually for the last ten years. In 1969 the mine output was approximately 500,000 short tons, but the total used was three times this amount. The balance was obtained from imports and secondary recovery (1).

Lead is one of the most widely used nonferrous metals in the manufacture of metal products, pigments, chemicals, and a variety of other items. In 1969 the manufacture of storage batteries accounted for some 40% and gasoline additives some 20%, of the U. S. total consumption of lead. Red lead and litharge pigments, ammunition, solder, cable covering, and caulking lead together utilized some 26% (1). Lead melts at a relatively low temperature (327 C) and easily volatilizes (at 490 C). If only a small proportion of the annual production is released into the environment, potentially toxic situations could arise.

CYCLES

A scheme of the ecological pathways by which environmental lead may reach and be taken in by man is given as Fig. 1. As far as is known, environmental lead is in the inorganic form. There is no evidence that a lead alkyl or "methyl" lead is synthesized in nature in the way that occurs with mercury. The organic lead compounds added to gasoline are changed to oxides in the combustion process, and then converted to halides by the "scavengers" that have also been added to fuel to remove lead from metallic engine parts. Less than 10% of the lead persists in the exhaust in the organic form. To the extent that man is directly exposed to organic lead compounds in the course of their manufacture or use, the possibility of their intake is important because of the higher toxicity of the organic forms.

DISTRIBUTION IN THE ENVIRONMENT

Because lead has been used almost from the time that man learned to extract and fashion it into metallic objects, it is difficult to speak of a "natural" or "normal" background. For similar reasons, "average" concentrations in food, water, or air have little meaning beyond the somewhat reassuring fact that they apparently have not changed in recent decades. Present day distributions in soil, glaciers, and waters are simply the result of a

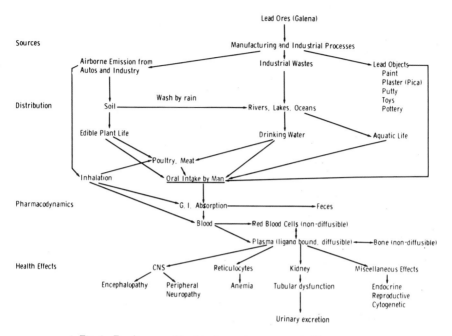

Fig. 1. Ecodiagram of lead in the environment and effects on man.

long continued and increasing addition to the environment, with an exponential surge following the industrial revolution.

Food

Food has long been a major source of human intake through contamination from lead-containing vessels (old pewter) or lead pottery glazes. More recently, pesticides have added their quota, and minor additional amounts may be derived from absorption of gasoline fumes in locations with high concentrations. Approximately half of the lead emitted by automobiles is deposited within 30m of the roadway (2). Roadside plants may contain up to 3000 ppm (3). Shellfish may concentrate lead from contaminated water. The average daily intake from food has been estimated as 300 μg for adults. Pica, the tendency of children and particularly of malnourished children to eat strange materials, often leads to the consumption of flaking paint, much of which may contain lead and give rise to chronic lead poisoning. Illicit liquor from stills contaminated with lead presents a similar opportunity to adults.

Air

In occupational situations where lead or lead-containing products are processed, lead may be found in significant quantities in the air, or be distributed on clothing or skin, and thus gain access to the body. Concentrations of lead in the air of industrial plants may reach 50 mg/m^3 (4), but good industrial hygiene can reduce this source of contamination and absorption to an acceptable minimum.

Effluents from smokestacks and other gaseous emissions from smelters and refining processes can distribute very significant quantities of lead as vapor, mist, or dust to the air and the surrounding soil, whence it can reach man through the ingestion of food or water or by inhalation.

Apart from local contaminations from industrial processes, or the burning of lead-containing wastes such as old battery cases, the most common source of lead contamination in ambient air today is automobile exhaust. 70 to 80% of the lead in gasoline is eventually discharged into the atmosphere. Concentrations range from 1 to 3 µg/m^3 in "average" urban air to 40 µg near heavy traffic. If, as has been estimated (5), some 25-40% of the inhaled lead is retained in the body, the contribution from air is of the same order as that from food. Neither of these are hazardous in themselves, but the air content has received considerable attention in recent years for good reasons: it is still rising; it is unnecessary and controllable; it adds yet another quota to the burden imposed upon people who are unduly susceptible to toxic effects, or whose exposure is approaching acceptable limits from some other source.

Water

In areas where limestone and galena ores are found, natural waters may contain lead in solution up to 0.8mg/li (6). Surface waters otherwise contain significant amounts of lead only when subject to some special contamination. In the United States, only 5

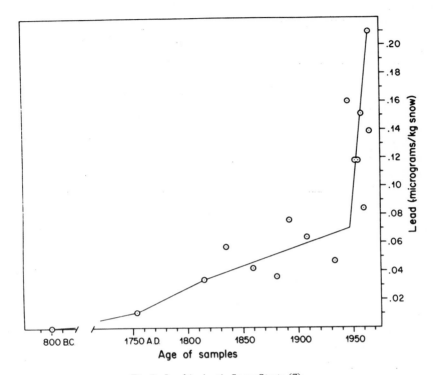

Fig. 2. Lead in Arctic Snow Strata (7).

out of 876 samples examined in 1962-4 contained more
than 50 µg/li. Approximately 14% of representative
drinking water supplies (i.e., bulk sources of piped
drinking water) in 1963-5 contained more than 10
µg/li, but less than 1% exceeded 30 µg/li. Rainwater
collected near a busy highway, on the other hand, may
contain as much as 50 mg/li (5). The lead content of
arctic snow is thought to reflect increasing
contamination of the atmosphere (7). The lead
content of snow deposited during the last century
shows a steep increase during the past 20 years,
whereas the content of other minerals has remained
steady (Fig. 2).

Additional lead may be added to drinking water
as it flows from the bulk supply. Lead chromate
paints have been used on the inside of steel storage
tanks. Old lead pipe installations still exist and
may permit lead to be absorbed, particularly by soft,
acid waters. Copper tubing, commonly used for the

61

distribution of water throughout a building, may be joined by lead solders. A good survey of the lead content of water as it comes out of the faucet seems not to have been made.

Soil

Lead reaching soil from water, or by deposition from the air, is likely to be fixed by the clay and converted to insoluble, inactive compounds (8). Very acid soils, however, may render the lead more soluble and thus more easily returned to the ground water.

TOXICOLOGY AND PHARMACOLOGY

ABSORPTION

In most instances, food and beverages provide the largest source of lead for man. The average daily lead intake of healthy adult persons in the United States varies from less than 0.10 mg/day to more than 2 mg/day (9). The mean intake of lead from food and beverage in persons with varied diets is approximately 0.30 mg/day.

Net absorption of lead from the gastrointestinal tract is 5-15% or less; the remaining lead is excreted in the feces. Experimental studies with dogs and rats indicate that a large percent of orally ingested lead is absorbed in the duodenum, transported through the liver and re-excreted into the gastrointestinal tract with bile (10,11).

The contribution of airborne lead to body lead content is dependent on many factors and is difficult to measure. The two most important factors with regard to the fate of the inhaled particles appear to be size and distribution in the respiratory tract. Most atmospheric lead is particulate lead halide from exhaust emission from automobiles. Lead particle size in urban air varies over a wide range (0.16-0.43μ) and has an average mass median diameter of 0.25 μ (12,13). A model for estimation of dust deposition in the respiratory tract has been proposed by the Task Group on Lung Dynamics (14). The respiratory tract may be described as made up of three functional areas: the nasopharynx (N-P), the tracheobronchial tree including the terminal bronchials (T-B) and a pulmonary compartment (P) consisting of respiratory bronchioles, alveolar ducts, atria, alveoli and alveolar sacs. Only a very minor fraction of particles under 0.5µ are retained in the (N-P) or (T-B) compartments. The remainder are cleared by ciliary action of respiratory epithelial cells and swallowed. The percentage of particles less than 0.5µ retained in the pulmonary compartment increases with reduction in particle size. The studies of Nozaki (15) on the influence of respiratory rate and particle size on retention of lead in the respiratory tract showed that 22-63% of particles 0.1 to 1.0 µ are deposited in the lung. In an experimental estimate of retention of inhaled particles (50% <0.9µ) performed on a human subject by Kehoe (16), it was found that about 45% of the particles were retained by the respiratory tract and about 40% of this fraction subsequently found its way into the gastrointestinal tract. The gut absorbs about 10 to 12% of this latter fraction. Retention by the pulmonary compartment is not equivalent to absorption, however. Recent experimental evidence indicates that a portion of such particles can be expected to be cleared by pulmonary macrophages (17).

Estimates of respiratory inhalation of airborne lead particles by an individual in downtown Cincinnati may be as high as 30 to 40 µg/day, whereas a person in a rural environment might inhale less than 1 µg/day (18).

EXCRETION

Renal excretion is by two routes, glomerular filtration and transtubular passage or secretion (19). Experimental studies suggest that when blood lead levels are within the normal range, most urinary lead is excreted by glomerular filtration (20). It is likely that as blood lead levels are increased, particularly to pathological levels, transtubular passage or secretion of lead increases. Urinary lead excretion in humans not exposed to excess lead averages 0.03 mg/day (9).

BODY BURDEN

The daily excretion of lead by way of feces, urine, sweat, and hair closely approximates daily ingestion, so that most people are virtually in lead balance. The elegant long-term studies by Kehoe (16) demonstrate that, in the absence of "excessive" exposure to lead, there is not likely to be any significant change in total body lead content. These studies in the living must, however, be reconciled with post-mortem data which indicate that, while soft tissue lead levels apparently remain low and constant, bone lead levels tend to increase with age. (See later)

The measurable quantities of lead in body tissues have been referred to as "body burden of lead". The significance of these lead stores to human health is not known. Absorbed lead is distributed in bone and various soft tissues and totals 100 to 400 mg in most adults (Table 1). The largest concentration of lead in persons with no abnormal exposure to lead is in the bone, where lead is bound in a non-diffusible form. Lead so bound seems not to be toxic. It is apparently the small "mobile" fraction located principally in the soft tissues that is associated with observed toxic effects. Bone lead must, however, be in equilibrium with blood and soft tissue lead.

Bone lead increases with age (25,26) and bones in adult males contain higher concentrations of lead than do bones of adult females (25). Soft tissue lead does not correspond well with bone lead content, but does increase through the second decade. In later years, soft tissue lead does not change in a predictable manner except perhaps for progressive increases in the aorta (25).

The relatively large liver and kidney lead content may be related to the excretory function of these two organs, whereas only small amounts of lead are present in tissues like muscle and brains.

Lead concentration of teeth increases from about 30 ppm in young people to about 90 ppm in persons over 50 years of age. Again, lead content of teeth is probably not an index of body burden after 40 years of age. Teeth may take up lead from the external environment by ion exchange. Lead is most concentrated toward the enamel surface and may reach 500 ppm in peripheral enamel in older age groups with a declining gradient toward the center of the tooth (27).

TABLE 1 -- LEAD CONTENT OF TISSUE FROM 15 PERSONS WITH NO
ABNORMAL EXPOSURE TO LEAD (CONTROLS) AND PERSONS DYING FROM
INORGANIC AND ORGANIC (tetraethyl) LEAD INTOXICATION*

Tissue	Controls (9)	Lead Intoxication	
		Inorganic (21,22)	Organic (21,23)
Bone	0.67-3.59	5.6 -17.6	2.9
Liver	0.04-0.28	1.8 - 8.0	2.35-3.4
Kidney	0.02-0.16	0.6 - 5.5	0.79
Spleen	0.01-0.07	1.13	0.29
Heart	0.04	0.2 - 0.8	0
Brain	0.01-0.09	0.24-1.2	0.74-1.9
Portions of Brain (24)			
Basal Ganglia		0.196	
Cortical Grey Matter		0.218	
Cortical White Matter		0.037	

* Values are mg/100g of wet tissue, range or single value,
from references cited.

Lead accumulates in hair, providing an easily
accessible tissue for estimating level of body burden
or exposure to lead (28).

There is a large body of literature regarding
the influence of environmental factors and metabolic
alterations on body retention of lead. Nevertheless,
considerably more knowledge of the metabolic behavior
of lead at the cellular level is required before the
influence of such factors on susceptibility to lead
toxicity are clearly understood. For example,
experimental studies have shown that in rats fed a
particular sub-toxic dosage of lead, lowering the
dietary calcium greatly increases the lead content of
blood and soft tissue as well as of bone (29). Iron
deficiency has a similar effect (30).

PATHOLOGY AND SITE OF ACTION

The most prominent adverse effects of lead
involve three organ systems: the nervous system, the
hematopoietic system, and the kidney.

Effects on the Nervous System

Nervous system effects of lead are both
structural and functional, involving the brain and
cerebellum as well as the spinal cord and motor and
sensory nerves leading to specific areas of the body.
The pathologic features of the central nervous system
effect of lead (lead encephalopathy) are varied, but

in severe cases the whole brain is swollen or edematous. The lead content of brain in even fatal cases of lead poisoning is relatively small, which supports the notion that neural tissue is very sensitive to the toxic effects of lead (Table 1). The most severe cellular damage is usually in the cerebellar cortex but may be equally severe in the cerebral cortex. Changes in the basal ganglia are less common. The pathogenesis of the cellular damage is uncertain but some authorities believe they result from an effect of lead on the metabolism or biochemistry of the brain. The observed abnormalities may be the result of a deficit in energy metabolism (31).

The hemodynamics of the brain are also impaired because of swelling of the endothelial cells lining small capillaries. This capillary injury may be attended by altered permeability of the vessel walls and so be largely responsible for the edema of the brain found in acute lead encephalopathy (32).

Electron microscopic study of the peripheral nervous system in experimental lead poisoning suggests that lead produces degenerative changes in supporting cells of peripheral nerves and, secondarily, in the myelin sheath of nerves (33,34). In addition, lead seems to interact with calcium at the motor end plate of motor nerves thus interfering with impulse transmission (35).

Effects on Red Blood Cells

Anemia occurs in lead poisoning probably from two basic causes, impairment of heme synthesis and increased rate of destruction of red blood cells. Heme is the oxygen carrying component of hemoglobin, produced during the development of the red blood cells. The chemistry of heme formation continues through several steps, some of which are located within the mitochondrion, the energy producing organelle of the cell, as shown in Figure 3 (36-38). Lead may act at several steps in this process but a well-defined defect is known to occur at an early step, involving the enzyme, delta-aminolevulinic acid dehydratase (d-ala-dehydratase), resulting in increased serum and urinary levels of delta aminolevulinic acid (D-ALA). A later but important action of lead on heme synthesis is in a final step where it impairs the incorporation of iron molecules into heme (39). The mechanism of this lead effect is yet unclear. Bone marrow activity of the enzyme, heme ferro-lyase, which is involved in the incorporation of iron into the heme molecule, is reduced in experimental lead poisoning (40). Also, ultrastructural study of reticulocytes has demonstrated increased non-heme iron within mitochondria (41). The pathway for the biosynthesis of heme in red blood cells has also been demonstrated in other cells, so that it is probable that lead exerts similar effects on the biosynthesis of heme and heme proteins in other organ systems in the body. It has recently been shown that the cytochrome oxidase content of renal mitochondria from rats with lead toxicity is decreased (42). However, this facet of the potential effects of lead has not been adequately studied.

Lead decreases the life span of red blood cells, another factor contributing to the anemia of lead poisoning, but the mechanism responsible for this effect is not well understood (38,43). Lead also interferes in vitro with some of the metabolic properties of the red blood cell, such as the ability of red blood cells to transport sodium and potassium ions into and out of the cells. The physiological implications of this lead effect are uncertain (44,45).

Renal Effects

The excretory function of the, kidney commits this organ to a prominent role in lead metabolism and susceptibility to adverse effects of lead. Reabsorption of small organic molecules such as amino acids, glucose, uric acid, citric acid, and phosphate is decreased (46,59). This abnormality is not the result of a specific action of lead on any one of these substances. It probably reflects a decrease in the energy production essential to the transport of these substances from the renal tubular lumen, across the tubular lining cells, into capillaries adjacent to these cells (47). The decrease in energy metabolism is thought to reflect a direct effect of lead on renal mitochondria, the intracellular organelles containing the enzymatic apparatus responsible for energy production (48,49).

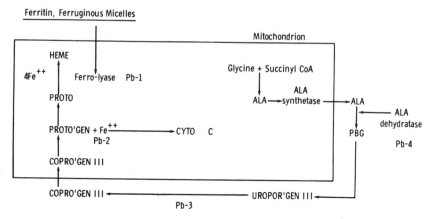

Fig. 3. Scheme of Heme Synthesis: Pb-1 inhibition of ferrochelatase; Pb-2 CPG oxidase, decarboxylase; Pb-3 UPG decarboxylase; Pb-4 d-ALA dehydratase. (Adapted from (36-38)).

A characteristic response of the kidneys of people and experimental animals exposed to large quantities of lead is the formation of intranuclear inclusion bodies in renal tubular lining cells (Figure 4). These bodies are composed of a lead-protein complex in which the lead is bound in a non-diffusible form (50). They are eventually excreted as the cells that contained them are shed into the urine (51). The major intracellular increment of lead during periods of increased lead ingestion is found in these lead-protein complexes (52). Since organelles like mitochondria are highly sensitive to the toxic effects of lead, it is hypothesized that the lead-containing inclusion bodies enable the kidney to excrete large amounts of lead without completely destroying the viability of tubular lining cells. Nevertheless, transport activity in the renal tubule is decreased in lead intoxication.

The formation of intranuclear inclusion bodies constitutes the most sensitive lead effect occurring in rats fed graded doses of lead in drinking water for a ten-week period. The bodies are recognizable by light microscopy in renal cell nuclei at a smaller dose of lead than that which will produce renal tubular dysfunction or elevated urinary D-ALA excretion (52). It has not been determined in humans, however, whether these bodies occur in the adaptive phase of lead poisoning, prior to the other

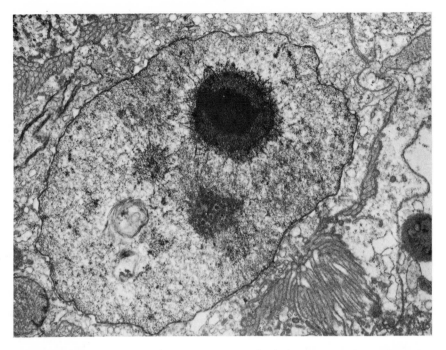

Fig. 4. Nucleus of a proximal renal tubular lining cell of lead poisoned rat containing an inclusion boyd (Pb incl) composed of a lead-protein complex, and a smaller incipient inclusion body (Sm incl). The lead-containing bodies are distinct from the nucleolus (Nucleo)(50)

manifestations of a lead effect. If so, recognition of these bodies in urine might serve as an early index of impending lead toxicity.

CLINICAL ASPECTS

HEALTH EFFECTS

Lead poisoning can give rise to several well-known but non-specific clinical syndromes of illness in man. Clinical experience suggests that each of these syndromes is associated with different intensities and lengths of excessive absorption of lead.

Anemia of Plumbism

An early clinical manifestation of lead poisoning is a mild anemia. Reticulocyte counts are moderately increased and basophilic stippling of the red blood cells in the peripheral circulation may also be seen. The peripheral red blood cells show the characteristic features of an iron deficiency anemia, which is not surprising in view of the evidence that lead can interfere with the utilization of iron for the formation of hemoglobin (37,38,41,53). The clinical features of mild anemia due to lead are indistinguishable from those of other chronic anemias due to a variety of causes. They include pallor, waxy sallow complexion, easy fatigability, irritability, and a mild headache. 'n young children this is frequently misinterpreted as a "mild behavior disturbance." Mild anemia may be the only clinical feature of prolonged moderately increased absorption of lead in which blood lead levels range between 60 and 120 µg Pb/100 g whole blood. Furthermore, anemia is virtually always present when the other more severe syndromes of acute lead poisoning are found.

Syndrome of Acute Abdominal Colic

In the adult, acute abdominal colic is usually preceded by one to two weeks with the onset of headache and usually generalized muscle aches. Following this, the patient becomes constipated, and within a few days attacks of cramping, diffuse, abdominal pains occur. When pains and constipation are severe, the patient begins to vomit. This may in turn be followed by loss of appetite, loss of weight, increasing fatigability, and usually a complaint of "bad taste in the mouth". Young children in the one to three year old range do not localize symptoms well, so at this stage illness may be expressed as anorexia, apathy, irritability, refusal to play, pugnaciousness, occasional vomiting, and constipation. In the absence of severe anemia, the syndrome of acute abdominal colic is almost always associated with blood lead levels in excess of 80 µg Pb/100g whole blood. Likewise, a careful appraisal nearly always reveals a recent increase in the absorption of lead. The symptoms subside over a period of several days or weeks when the abnormal exposure is terminated.

Syndrome of Acute Encephalopathy

This is the most severe acute clinical form of lead poisoning. It is usually associated with blood lead levels in excess of 120 μg Pb/100g. In children with encephalopathy, blood lead values ranging between 100 and 1310 μg Pb/100 g whole blood have been reported. Acute encephalopathy may present precipitously without warning, with the onset of intractable convulsions, lapse into coma, cardiorespiratory arrest, and death. There may be no history of any prodromal manifestations, and death may occur within 48 hours after the initial convulsion unless life is prolonged in the vegetative state through artificial cardio-pulmonary assistance. This brief catastrophic sequence may also occur against a background of anemia and mild colic. Most frequently, however, the fulminant form of acute encephalopathy develops during a period of approximately one week. In children between one to three years of age vomiting, which may have been sporadic, becomes increasingly frequent, persistent, and forceful. The child's apathy and unwillingness to play progress to drowsiness and stupor, interspersed with alternating lucid and hyperirritable intervals. During the final 48 hours these alterations in the state of consciousness progress rapidly to coma and convulsions. In young children careful history may reveal mental regression in the form of loss of recently acquired motor and verbal skills. About 20% of the patients have a history of recent onset of clumsiness. Such children usually show frank ataxia on examination. Both the onset and clinical course of encephalopathy are unpredictable. Symptoms may abate at any point in the above sequence if exposure is temporarily interrupted, but they will recur if abnormal exposure recurs. Prior to the availability of chelating agents for treatment, blood lead levels in excess of 500 μg Pb/100 g were usually associated with a fatal outcome. Acute encephalopathy may be accompanied by the Fanconi syndrome (generalized renal aminoaciduria, mellituria, and hypophosphatemia with hyperphosphaturia) (46). In adults, other alterations in renal function have also been reported (54). The minimum period of grossly excessive intake required to produce acute encephalopathy is estimated at approximately one month (55). The more usual minimal period is estimated at approximately three to

four months (56). Summer is apparently an important
precipitating factor although the mechanisms
responsible for this clinical observation are not
fully understood (57).

Syndrome of "Chronic Encephalopathy"

Occasionally children are seen in whom a
clear-cut history of progressive mental deterioration
is obtained. They are usually over three years of
age. The clinical history indicates normal
developmental progress during the first 12 to 18
months of life or longer, followed by a steady loss
of the skills acquired, the onset of a convulsive
disorder, and loss of speech to the point of mutism.
At the time of examination these children show severe
hyperkinetic and aggressive behavior disorders, they
may have a poorly controlled convulsive disorder, and
there may be evidence of current excessive absorption
of lead. Blood lead levels in excess of 60 μg
Pb/100g are usually found, and X-rays of the long
bones show heavy multiple bands of increased density
at the metaphyses of the growing long bones, which
indicate long continued excessive absorption of lead.
This clinical entity has been termed "chronic
encephalopathy". Although no history clearly
suggestive of prior episodes of acute encephalopathy
is obtained, this possibility can never be entirely
excluded. Indeed, the clinical picture of seizure
disorder, mental incompetence, and hyperkinetic
aggressive behavior disorder is the well recognized
sequel of acute lead encephalopathy during early
childhood and is especially likely to follow
recurrent episodes of acute encephalopathy (57-59).

Syndrome of Peripheral Neuropathy

The distinguishing clinical features of the
peripheral neuropathy of lead poisoning are
predominance of motor involvement, with minimal or
absent sensory abnormalities and a tendency to
involve the extensor muscles of the hand and feet.
Classically, wrist drop is said to be more prevalent
in adults and foot drop more common in children. It
is rarely recognized in children under three years of
age.

Syndrome of Late or Chronic Lead Nephropathy

Lead nephropathy is characterized by progressive and apparently irreversible renal insufficiency. Renal biopsies and post-mortem specimens show a non-specific picture of interstitial fibrosis, tubular degeneration, and vascular changes in small arteries and arterioles (60,61). Functionally and clinically, the distinguishing feature of lead nephropathy is a selective impairment in the excretion of uric acid, which results in a secondary hyperuricemia with or without manifest gout. Renal insufficiency with or without hypertension normally precedes the clinical onset of gout. Nye (62) was the first to note an association between chronic lead poisoning during childhood and the delayed onset of chronic lead nephropathy. Tepper (63), in a follow-up of children in the United States who had had lead poisoning during early childhood, could find no evidence of late renal injury. The difference between these two groups apparently lies in the length of exposure. The patients studied by Tepper (63) had apparently been exposed to excessive amounts of lead during the pre-school years only; whereas, those studied by Nye (62) and more recently by Emmerson (59) were apparently absorbing excessive amounts of lead over a much longer period of time. The same syndrome has also been found in long-term imbibers of lead-contaminated "moonshine" whiskey (60). Clinical evidence indicates, therefore, that long-term excessive exposure, perhaps for a period of approximately ten years or longer, is necessary to produce late lead nephropathy. Many investigators are not satisfied that lead per se is entirely responsible for this syndrome. Certainly the available studies do not exclude the possibility that other unrecognized factors may play a contributing role in the clinical picture of "chronic lead nephropathy".

Sequelae in Children

At least 25% of the young children who survive an attack of acute encephalopathy sustain severe permanent brain damage (64). Chisolm and Harrison (56) reported that the return of a child following a single known attack of acute encephalopathy to the same hazardous home environment increases the risk of severe permanent brain damage in that child to virtually 100%. Subtle neurologic deficits and mental impairment are the more common outcome. These include lack of sensory perception and perseveration, despite I.Q. scores of 80 to 100 or better on verbally oriented intelligence tests. The sense of form and proportion is distorted. Motor incoordination and lack of sensory perception severely impair the child's learning ability. Often the handicap is not recognized until after the child enters school. Such children also have short attention span and easy distractibility. Many lead-poisoned children develop hostile, aggressive, and destructive behavior patterns which, in turn, may precipitate their exclusion from school and the demand for institutionalization. Although seizure disorder and behavioral abnormalities tend to abate during adolescence, mental incompetence is permanent (59,65).

Perlstein and Attala (57) studied 425 children followed six months to ten years after the initial diagnosis of lead poisoning or asymptomatic increased lead absorption. Their findings are summarized in Table 2.

Similarly, Byers and Lord (58) noted permanent mental subnormality in 19 out of 20 children who did not have clear-cut encephalopathy but who sustained recurrent bouts of acute plumbism during the pre-school years. Because the true incidence of lead poisoning in young children in inner cities is not known, the incidence of significant permanent injury to the central nervous system is also not known. The data cited above suggest, however, that it is both a significant and a preventable cause of brain damage in children. Whether asymptomatic increased lead absorption can cause subtle but permanent impairment of nervous system function in young children is not known.

Sequelae in Adults

The older medical literature refers to similar mental impairment, encephalopathy, and seizures associated with uncontrolled occupational exposure to lead. In the past this was associated during the acute episode with blood lead levels well above 100

TABLE 2 -- DISTRIBUTION OF SEQUELAE FOLLOWING VARIOUS MODES OF
ONSET IN 425 PATIENTS WITH PLUMBISM (57)

Sequelae	Mode of Onset						
	Enceph.	Seizures	Ataxia	G.I.	Febrile	Asymptomatic	Total
None	11 (18)	14 (33)	7 (41)	159 (69)	13 (81)	53 (91)	257 (61)
Mental retardation	23 (38)	14 (33)	5 (29)	43 (19)	3 (19)	5 (9)	93 (22)
Seizures	32 (54)	17 (39)	6 (35)	30 (13)	0	0	85 (20)
Cerebral palsy	8 (13)	0	1 (6)	0	0	0	9 (2)
Optic atrophy	4 (6)	0	1 (6)	0	0	0	5 (1)
All cases	59	43	17	232	16	58	425

Data in parentheses are percentages of cases with that mode of onset developing the sequelae.
The percentages in any column may total more than 100 because the one patient may develop
more than one sequela.

µg Pb/100 g whole blood. Such levels are uncommon in occupational exposure today. Similarly, the occurrence of severe plumbism is uncommon. In the adult, the long-term sequelae of lead poisoning seen today include peripheral neuropathy and lead nephropathy as discussed previously, these being associated apparently with prolonged excessive exposure to lead and a history of recurrent episodes of acute poisoning.

Other Effects

ENDOCRINE EFFECTS -- Recent clinical studies in persons with chronic lead poisoning from ingestion of lead contaminated whiskey have been shown to have diminished secretion of pituitary gonadotropins and hypothyroidism. Further study of lead poisoned rats suggests that lead impairs uptake of iodine by the thyroid gland as well as decreasing the rate of protein binding of iodine (66).

REPRODUCTIVE EFFECTS -- Severe lead intoxication has been associated with sterility, abortion, still births, and neonatal deaths in man (67-69). Modern documentation of lead intoxication as a cause of injury to the fetus is not clear.

PULMONARY EFFECTS -- Possible adverse effects of lead on the lung are of interest since one source of exposure to environmental lead is from lead aerosols produced by auto and industrial emission. Recent experiments have shown that inhalation by rats of an aerosol containing lead particles less than 0.1µ reduces the number of alveolar macrophages in pulmonary washings (17). The implication is that lead inhalation may decrease the pulmonary defense to other particulate contaminants and infectious agents.

LEAD AND CANCER -- Chronic lead toxicity in rats and mice results in renal cancer, (70,71) and dietary lead fed with 2-acetylaminofluorene will produce cerebral gliomas in rats (72). There is no evidence that lead is carcinogenic in man.

DOSE-RESPONSE RELATIONSHIPS

In man, the effects of lead on the biosynthesis of heme and on the function of the kidney, the nervous system and the hematopoietic system may be related in an approximate way to differing levels of absorption of lead. Perhaps, the best direct index of current and recent absorption of inorganic lead (salts) is provided by serial measurements of the concentration of lead in whole blood.

Whole blood lead levels in the healthy population vary from about 15 to 40 µg/100g. Blood lead levels may be correlated with environmental exposure, but, nevertheless, are remarkably constant in various parts of the world (Table 3). The blood lead level of New Guinea aborigines is greater than that found in residents of suburban Philadelphia and the blood lead level of rural Peruvians is only slightly lower than that of suburban North Americans. These data suggest that man may be exposed to different (non-gasoline) sources of environmental lead in other parts of the world.

With few exceptions, mean blood lead levels in groups of healthy adults are in the neighborhood of 20 µg Pb/100g. In one international study (18), the aggregate mean blood lead level was 22 µg Pb/100g (S.D. ± 5). These data show that the vast majority of the general population fall within the range of "no demonstrable in vivo effect" as described in Table 4. This appears to be the level of absorption associated with traces of lead in usual dietary sources and most respiratory exposures to lead in the ambient air.

TABLE 3 -- INFLUENCE OF ENVIRONMENTAL AND GEOGRAPHIC
LOCATION ON BLOOD LEAD LEVELS IN ADULTS

A. Occupations in U. S. A. (73)	µg/100 ml
Suburban non smokers, Philadelphia	11
Suburban smokers, Philadelphia	15
All Policemen, Cincinnati	25
Service Station attendants, Cincinnati	28
Traffic Police, Cincinnati	30
Tunnel employees, Boston	30
Parking lot attendants, Cincinnati	34
Garage mechanics, Cincinnati	38

B. Geographic Differences (74)	
Mean, Samples 14 countries	17
Lowest mean, Peru	7
Highest mean, Helsinki, Finland	26
New Guinea Aborigines	22

TABLE 4 -- EFFECTS OF INORGANIC LEAD SALTS IN RELATION TO ABSORPTION

(a) Classification of Levels and Types of Effects

Types of Effects:	1. No demonstrable in vivo effect	2. Minimal Subclinical metabolic effect	3. Compensatory biologic mechanisms invoked (Mild)	4. Acute lead poisoning (Severe)	(Severe)	5. Late effects of chronic or recurrent acute lead poisoning
A. Metabolic: (accumulation and excretion of heme precursors)	normal	slight increase in urinary ALA may be present	progressive increase in ALA, UCP, FEP	Increased 5 to 100 fold		Increased if excessive exposure recent, but may not be increased if excessive exposure remote
B. Functional Injury: 1. Hematopoiesis	none	none	shortened RBC life span, reticulocytosis (+) (reversible)	shortened RBC life span and reticulocytosis with or without anemia (reversible)		Anemia (+) (reversible)
2. Kidney (renal tubular function)	none	none	?	aminoaciduria, glycosuria (+) (reversible)	Fanconi syndrome (reversible)	chronic nephropathy (permanent)
3. Central nervous System	none	none	?	mild injury (??? reversible)	severe injury (permanent)	severe injury (permanent)
4. Peripheral nerves	none	none	?	rare	rare	Impaired conduction (wrist, foot drop usually improve slowly, but may be permanent)

C. Clinical Effects:

			non-specific mild symptoms (may be due in part to co-existing diseases)	colic, irritability, vomiting	ataxia, stupor, coma, convulsions	mental deficiency (may be profound) seizure disorder, renal insufficiency (gout) (permanent)
none	none	none				

(b) Index of Recent or Current Absorption by Level of Effect

						may be normal
Blood Level (μg Pb/100 gm whole blood)	<40 μg Pb	40–60 μg Pb	Level III. 60-----100+ μg Pb	Level IV with anemia, intercurrent disease. 60---------100 + μg Pb / Level IV. >80 μg Pb		
Urine Lead (adults only) (μg Pb/L)	<80 μg Pb/L	<130 μg Pb/L		>130 μg Pb/L (may be less in severe illness)	endogenous excretion may be normal	CaEDTA mobilization test: chronic nephropathy-positive permanent CNS injury ± (see text)

Pb = lead; ALAD = δ-aminolevulinic acid dehydratase; ALA = δ-aminolevulinic acid; UCP = urinary coproporphyrin; FEP = free erythrocyte protoporphyrin; RBC = red blood cell. See text and Fig. 5 for relationship between blood lead level and activity of ALAD as measured in vitro in hemolysates of blood.

The "Tri-cities" study (4) indicated that mean blood lead levels are slightly higher (25-38 μg Pb/100g) in some groups of men whose occupation brings them into close daily contact with motor vehicular traffic (e.g., traffic policemen, tunnel employees, garage attendants). However, in no instance did the blood lead level exceed 43 μg Pb/100g in this study (75). In contrast, a study of inner city children revealed a mean blood lead level of 43 μg Pb/100g (S.D. ± 11 μg Pb) (76). In another study in the same city, the average fecal output of lead was 650 μg Pb/day in a small number of children without known pica who lived in old deteriorated urban housing (56). These data indicate that a large proportion of these individuals may be expected to show at least a minimal increase in the urinary excretion of ALA, but that they would be unlikely to show any recognized adverse functional effect (Level II and III, Table 4).

Some years ago, Kehoe (16) fed supplemental amounts of lead (0.3, 1.0, 2.0, and 3.0 mg Pb/day) as soluble inorganic lead salts to human adult volunteers in addition to the normal daily dietary intake (0.3 mg Pb on the average). The men who received 1.0, 2.0, and 3.0 mg Pb/day showed a progressive rise in blood lead level from the "normal" level of approximately 30 μg Pb/100g whole blood to a range of 50 to 70 μg. The rate of increase was faster in the men fed the higher doses of lead. None of these health volunteers developed symptoms of poisoning although, at the doses administered, the rises in blood lead concentration occurred over periods of several months to several years. When increased intake was halted, blood lead levels decreased. The blood lead concentration reflects recent and current absorption of lead; the various adverse effects of lead are associated with progressive increments in the levels of lead in whole blood. The metabolic effects of lead on heme synthesis and hematopoiesis are reversible upon termination of excessive exposure; however, some of the functional injury to the kidney and the nervous system associated with acute symptomatic attacks of plumbism (lead poisoning) may not be fully reversible. In general, the more severe the illness, the higher the blood lead level, and the more prolonged the abnormal level of absorption, the greater is the risk of permanent injury to the kidney and nervous system.

The various effects of lead are shown schematically in Table 4, together with the different ranges of blood lead levels associated with each type of effect. Each level of effect may be discussed with respect to the significance of each type of effect on human health. In so doing, one draws heavily upon experimental data and past and present clinical and epidemiological experience. This is necessary because the actual dosage of lead can rarely be determined in the clinical situation.

Normal or "Natural" (Level I)

All persons have very small but measurable quantities of lead in blood, urine and tissues. Although no essential requirement for the metal has been demonstrated, no metabolic change is demonstrable in vivo in persons with blood lead concentrations of less than approximately 40 μg Pb/100g. However, a negative correlation has been found between blood lead concentration and the activity of δ-aminolevulinic acid dehydratase (ALAD) in hemolysates of blood (Figure 5) when assayed in vitro (77-81). The physiologic significance of this in vitro observation (inhibition of ALAD activity in circulating red blood cells) is uncertain: Hernberg and others (78,82) suggest that the assay provides a measure of exposure to lead (as does the blood lead level).

The metabolic consequence of inhibition of ALAD activity in vivo in the various organs of the body is accumulation and increased excretion of the unutilized portion of the enzyme's substrate; namely, δ-aminolevulinic acid (ALA). The date of Selander and Cramer (83) in Figure 6 indicate that the first detectable increase in the concentration of ALA in urine can be seen as blood lead concentration increases from 40 to 60 μg Pb/100g (Level II in Table 4). Even so, no impairment of organ function attributable to lead has been demonstrated in otherwise healthy adults with these findings (Levels I and II in Table 4).

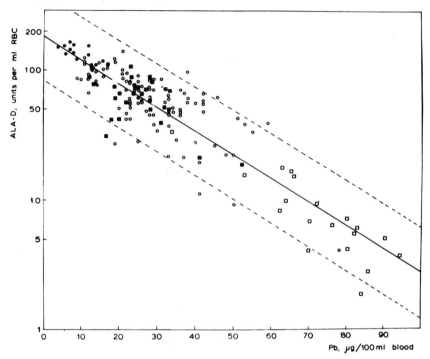

Fig. 5. Correlation between ALAD and Pb in blood of 158 persons representing different degrees of natural and occupational exposure to lead. Note logarithmic scale on ordinate. Solid circle medical students; open circle workers in printing shops; solid square automobile repair workers; open square lead smelters and shipscrappers. (78)

Increased Lead Absorption

As blood lead levels increase beyond approximately 60 µg Pb/100g, variation in the responses of individuals to increasing levels of lead in blood becomes evident. Some individuals show twofold or greater increments in the urinary concentration of ALA while others do not (Figure 6). On the other hand, virtually all persons with blood lead concentrations 80 to 90 µg Pb/100g show increased amounts of ALA in serum (36,84,85) and urine (36,78,83) as in the example shown in Figure 6.

Similar responses in the urinary excretion of lead and coproporphyrin have also been reported (36,83). Other in vitro metabolic effects may also become demonstrable as blood lead level concentrations exceed 50 µg/100g. No clear-cut evidence of

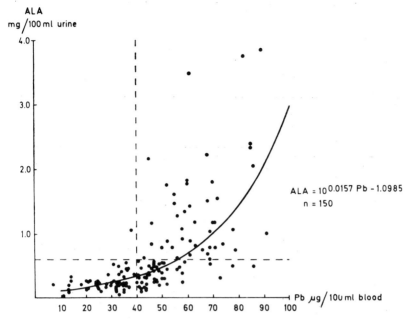

Fig. 6. Values for LAA in urine plotted against those for lead in blood. Dotted lines mark upper normal limits (83)

functional impairment of the red cells is found at this stage. This suggests that compensatory biologic mechanisms may be playing a modifying role in response to the increased absorption of lead (Level III in Table 4). Reduced red blood cell survival is correlated with increasing reticulocytosis (86) which suggests that the compensatory response of the hematopoietic system to increasing levels of lead is an increased rate in the production of new erythrocytes. Anemia results when this compensation fails. Reduction in red cell life span to <105 days appears to be related to blood lead levels >80-100 µg Pb/100g (81). These and other aspects of a complex nature of the anemia of lead poisoning have been reviewed by Griggs (37), Waldron (38), Bessis and Jensen (41), and Sana (52).

At this level of absorption and compensatory metabolic response, no clear-cut laboratory evidence of functional impairment of kidney and nervous system may be found. Symptoms, if present, are non-specific (headache, abdominal discomfort and "not up to par") and may be due, in toto or in part, to other intercurrent illnesses which may also be present. In Table 4, a rather wide overlap in blood lead values may be noted between Levels III and IV: Thus, clinical signs compatible with plumbism may be seen in some persons with blood lead levels as low as 60 μg Pb/100g, while others may be asymptomatic at levels of 100 μg Pb or slightly higher (36,38,81,87). Other unrelated disease states (especially those associated with iron deficiency) are usually present when obvious clinical plumbism occurs in the presence of lower levels of lead in the blood.

Blood levels associated with frank clinical effects were given above in describing health effects. Clinical data indicate that the levels of intake associated with symptoms of illness and demonstrable functional injury are 10-250 times greater than the levels of intake to which the vast majority of the general population is now exposed. Clinical experience further suggests that irreversible injury is associated with either severe acute illness (encephalopathy) or long-term high levels intake and recurrent episodes of acute lead poisoning (56-61,88). Whether asymptomatic increased lead absorption is associated with significant and deleterious functional consequences, especially in young children, is not known. No data pertinent to this particular problem are available, nor can they be deduced from clinical experience since past intermittent episodes of mild acute poisoning can never be entirely excluded in the reported clinical studies.

DIAGNOSTIC INDICES

Older medical texts refer to basophilic stippling of red cells and "lead lines" on the gums as the hallmarks of lead poisoning in humans. It is now recognized that these clinical features are the hallmarks of very prolonged and uncontrolled excessive absorption of lead. They are rarely encountered today. As described above and mentioned earlier, the clinical signs and symptoms of lead poisoning are usually non-specific and may be produced by a variety of other diseases. For this reason, clinical diagnosis depends upon the knowledge that the patient has had some unusual occupational or non-occupational exposure to lead and the performance of certain specialized laboratory tests. These tests include measures of the absorption of lead, such as blood lead concentration and, in adults but not in young children, urinary lead excretion. X-rays of the long bones in young children but not in adults may reveal excessive storage of lead in the form of bands of increased density at the metaphysis of the long bones. In addition, the adverse metabolic effects of lead must be demonstrated in the form of increased output of ALA and coproporphyrin in the urine, increased free erythrocyte protoporphyrin, and increased concentration of ALA in serum (36,85). The normal adult levels of the metabolic precursors of heme which are raised in lead poisoning are: free erythrocyte protoporphyrin, less than 50 µg/100 ml packed red blood cells (53); plasma ALA, less than 10 µg/100 ml (85) or less than 20 µg/100 ml (36,84), depending on the method of analysis used; urinary ALA, less than 3.5 mgm/24 hr in adults (36) and less than 0.08 mgm/kg body weight/24 hr in children (83); coproporphyrin in urine, less than 175 µg/24 hours in adults (90), and less than 2 µg/kg body weight/24 hr in children (91). As blood lead concentration rises above 80µg Pb/100g whole blood, all of these metabolic indices of inhibition of heme synthesis become raised and in obvious clinical cases of lead poisoning they may be raised 5 to 100 fold above normal.

Fecal output of lead provides the best estimate of usual dietary intake, since approximately 90% or more of ingested lead passes through the intestinal tract and is excreted in the feces (16). Healthy adults without undue occupational exposure excrete approximately 300 µg Pb/24 hr on the average, while young children without pica who live in well-maintained housing excrete approximately 130 µg Pb/24 hr (56,92).

In young children with pica who consume leaded house paint, a mean fecal output of approximately 50 mg Pb/24 hr was found in 13 children with acute symptomatic lead poisoning (56). In five children with asymptomatic increased lead absorption, average daily fecal output of lead was approximately 2 mg/24 hr (56). The very wide range in the daily fecal output of lead in these children (0.3-252 mg Pb/24 hr) probably reflects the episodic nature of pica.

In children with asymptomatic increased lead absorption, the fecal data suggest that the level of intake is about ten times greater than that found in normal dietary intake. The amounts of lead absorbed under these conditions can only be approximated; however, children with lead encephalopathy have been found to excrete about 37 mg Pb in the urine during a five-day course of chelation therapy with BAL and EDTA (85). A second course of chelation therapy usually evokes a further excretion of approximately 10-15 mg Pb. This excretion is equivalent to 3-5 mg Pb/kg body weight and represents that fraction of the total lead burden accessible to this combination of chelating agents.

RELATIVE RISK

The vast majority of the population is, with few exceptions, not experiencing levels of absorption of lead which are thought to have any effect upon their health. A few people in congested urban areas may be expected to show a minimal increase in urinary ALA excretion only. Any further increment in general lead exposure, particularly in urban areas, would be expected to bring a somewhat larger proportion of people into this range. Today in the United States, however, industrial workers in the lead trades, young children in deteriorated housing in urban areas, and imbibers of moonshine whiskey are the groups principally at risk for the subclinical and clinical adverse effects of lead. For these groups, increments in overall environmental contamination by lead can only add to their risk.

REFERENCES

1. Lutz G. A. et al. (1970). Final report on technical, intelligence, and project informational system. 3. Lead model case study. Battelle Memorial Institute.

2. Singer, M. and Hanson, L. (1969). Lead accumuation im soils near highways in the twin cities metropolitan area. Soil Sci. Soc. Amer. Proc. 33: 152.

3. Cannon, H. and Bowles, J. (1962). Contamination of vegetation by tetraethyl lead. Science 137: 765.

4. Working Group on Lead Contamination. (1970). NAPCA - Survey of lead in the atmosphere of three urban communities. Public Health Service Publication 999-AP-12, U.S. GPO.

5. Symposium on Environmental Lead Contamination. (1966). U.S. Public Health Service Publication 1440.

6. Kopp, J. F. and Kroner, R. C. (1967). A five-year summary of trace metals in rivers and lakes of the United States (1962-1967). Federal Water Pollution Control Administration.

7. Murozumi, M., Chow, T. J. and Patterson, C. (1969). Chemical concentrations of pollutant lead aerosols, terrestrial dusts and sea salts in Greenland and Antarctic snow strata. Geochim. Cosmochim. Acta 33: 1247.

8. Gilber, F. A. (1957). Mineral metabolism and the balance of life. University of Oklahoma Press, Oklahoma City, Oklahoma.

9. Kehoe, R. A. (1961). The metabolism of lead in health and disease: Present hygenic problems relating to the absorption of lead. J. Royal Inst. of Pub. Health and Hyg. 24: 177.

10. Black, S. C. (1962). Storage and excretion of lead 210 in dogs. Arch. Env. Health 5: 423.

11. Castellino, N. and Aloj, S. (1964). Kinetics of the distribution and excretion of lead in the rat. Brit. J. Ind. Med. 21: 308.

12. Robinson, E. and Ludwig, F. L. (1967). Particle size distribution of urban lead aerosols. J. Air Pollut. Contr. Assoc. 17: 664.

13. Lee, R. E. (1968). Particle size distribution of metal components in urban air. Env. Sci. Tech. 2: 288.

14. Task Groups on Lung Dynamics. (1966). Deposition and retention models for internal dosimetry of the human respiratory tract. Health Physics 12: 173.

15. Nozaki, K. (1966). Method for studies on inhaled particles in human respiratory system and retention of lead fume. Ind. Health 4: 118.

16. Kehoe, R. A. (1961). The metabolism of lead in man in health and disease: The metabolism of lead under abnormal conditions. J. Roy. Inst. Pub. Health and Hyg. 24: 101.

17. Bingham, E., et al. (1968). Alveolar macrophages: Reduced number in rats after prolonged inhalation of lead sesquioxide. Science 162: 1297.

18. Goldsmith, J. R. and Hexter, A. C. (1967). Respiratory exposure to lead: Epidemiological and experimental dose-response relationships. Science 158: 132.

19. Vostal, J. and Heller, J. (1968). Renal excretory mechanisms of heavy metals. 1. Transtubular transport of heavy metal ions in the avian kidney. Env. Res. 2: 1.

20. Teisinger, J. (1966). Relationship between the lead content of blood and urine in subjects not exposed to lead. Cas. Lek. Ces. (J. of Czech Physicians) 105: 810.

21. Cumings, J. N. (1967). Trace metals of the brain and their importance in human disease. Chemisch Weekblad 63: 473.

22. Theirs, C. H. and Haley, T. J. (1964). Clinical Toxicology. (4th ed.). Lea and Febiger, Philadelphia, p. 141.

23. Cassels, D. A. K. and Dodds, E. C. (1946). Tetraethyl poisoning. Brit. Med. J. 2: 681.

24. Klein, M., et al. (1970). Earthenware containers as a source of fatal lead poisoning. New Eng. J. Med. 283: 669.

25. Schroeder, H. A. and Tipton, I. H. (1968). The human body burden of lead. Arch. Env. Health 17: 965.

26. Barry, P. S. I. and Mossman, D. B. (1970). Lead concentrations in human tissues. Brit. J. Ind. Med. 27: 339.

27. Stoehlow, C. D. and Kneip, T. J. (1969). The distribution of lead and zinc in the human skeleton. Am. Ind. Hyg. Assoc. J. 30: 372.

28. Kopito, L., Byers, R. K. and Shwachman, H. (1967). Lead in hair of children with chronic lead poisoning. New Eng. J. Med. 276: 949.

29. Six, K. M. and Goyer, R. A. (1970). Experimental enhancement of lead toxicity by low dietary calcium. J. Lab. Clin. Med. 76: 933.

30. Six, K. M. and Goyer, R. A. (1971). The influence of iron deficiency on tissue content and toxicity of ingested lead in the rat. (manuscript)

31. Pentschew, A. (1965). Morphology and morphogenesis of lead encephalopathy. Acta Neuropath. 5: 133.

32. Popoff, N., Weinberg, S. and Feigin, I. (1963). Pathologic observations in lead encephalopathy with special reference to the vascular changes. Neurology 13: 101.

33. Okazaki, H., et al. (1963). Acute lead encephalopathy of childhood. Histologic and chemical studies with particular reference to the angiopathic aspects. Trans. Am. Neurol. Assoc. 88: 248.

34. Lampert, P. W. (1968). Demyelination and remyelination of lead neuropathy. Electron microscopic studies. J. Neuropath. Exp. Neurol. 27: 527.

35. Kostial, K. and Vouk, V. (1957). Lead ions and synaptic transmission in the superior

cervical ganglion of the cat. Brit. J. Pharmacol. 12: 219.

36. Haeger-Aronsen, B. (1960). Studies on urinary excretion of amino-levulinic acid and other heme precursors in lead workers and lead-intoxicated rabbits. Scand. J. Clin. Lab. Invest. 12: Sup. 47.

37. Griggs, R. C. (1964). Lead poisoning: Hematologic aspects. In: E. B. Brown, and C. V. Moore (eds.). Progress in Hematology. Grune and Stratton, New York. p. 117.

38. Waldron, H. A. (1966). The anameia of lead poisoning: a review. Brit. J. Ind. Med. 23: 83.

39. Jandl, J. H. et al. (1959). Transfer of iron from serum iron-binding protein to human reticulocytes. J. Clin. Invest. 38: 161.

40. Otrzonsek, N. (1967). The activity of heme ferro-lyase in rat and bone marrow in experimental lead poisoning. Internat. Arch. Gewerbepath. Gewerbehyg. 24: 60,66.

41. Bessis, M. C. and Jensen, W. N. (1965). Sideroblastic anaemia, mitochondria, and erythroblastic iron. Brit. J. Haemat. 11: 49.

42. Rhyne, B. and Goyer, R. A. (1971). Cytochrome content of kidney mitochondria in experimental lead poisoning. Exptl. Molec. Pathol. 76: 933.

43. Berk, P. D. et al. (1970). Hematologic and biochemical studies in a case of lead poisoning. Am. J. Med. 48: 137.

44. Hasan, J. et al. (1967). Enhanced potassium loss in blood cells from men exposed to lead. Arch. Env. Health 14: 309.

45. Hasan, J., Vihko, V. and Hernberg, S. (1967). Deficient red cell membrane $Na^+ + K^+$-ATPase in lead poisoning. Arch. Env. Health 14: 313.

46. Chisolm, J. J., Jr. (1962). Aminoaciduria as a manifestation of renal tubular injury in lead intoxication and a comparison with patterns of amino-aciduria seen in other diseases. J. Pediat. 60: 1.

47. Goyer, R. A. (1968). The renal tubule in lead poisoning. 1. Mitochondrial swelling and aminoaciduria. Lab. Invest. 19: 71.

48. Goyer, R. A., Krall, A. and Kimball, J. P. (1968). The renal tubule in lead poisoning. 2. In vitro studies of mitochondrial structure and function. Lab. Invest. 19: 78.

49. Goyer, R. A. and Krall, R. C. (1969). Ultrastructural transformation in mitochondria isolated from kidneys of normal and lead intoxicated rats. J. Cell Biol. 41: 393.

50. Goyer, R. A. et al. (1970). Lead and protein content of isolated inclusion bodies from kidneys of lead-poisoned rats. Lab. Invest. 22: 245.

51. Landing, B. and Nakai, H. (1959). Histochemical properties of renal lead-inclusions and their demonstration in urinary sediment. Am. J. Clin. Path. 32: 499.

52. Goyer, R. A. et al. (1970). Lead dosage and the role of the intranuclear inclusion body. Arch. Env. Health 20: 705.

53. Sana, S. (1958). The effect of mitochondria on porphyrin and heme biosynthesis in red blood cells. Acta Haemat. Jap. 21: Sup. 2.

54. Sandstead, H. H., Michelakis, A. M. and Temple, T. E. (1970). Lead intoxication, its effect on the renin aldosterone response to sodium deprivation. Arch. Env. Health 20: 356.

55. Klein, M., et al. (1970). Earthenware containers as a source of fatal lead poisoning. New Eng. J. Med. 283: 669.

56. Chisolm, J. J., Jr. and Harrison, H. E. (1956). The exposure of children to lead. Pediatrics 18: 943.

57. Perlstein, M. A. and Attala, R. (1966). Neurologic sequelae of plumbism in children. Clin. Ped. 5: 292.

58. Byers, R. K. and Lord, E. E. (1943). Late effects of lead poisoning on mental development. Am. J. Dis. Child. 66: 471.

59. Emmerson, B. T. (1968). The clinical differentiation of lead gout from primary gout. Arthritis Rheum. 11: 632.

60. Morgan, J. M., Hartley, M. W. and Miller, R. E. (1966). Nephropathy in chronic lead poisoning. Arch. Intern. Med. 118: 17.

61. Galle, P. and Morel-Maroger, L. (1965). Les lesions renales du saturnisme humain et experimental. Nephron 2: 273.

62. Nye, L. J. J. (1929). An investigation of the extraordinary incidence of chronic nephritis in young people in Queensland. Med. J. Aust. 2: 145.

63. Tepper, L. B. (1963). Renal function subsequent to childhood lead poisoning. Arch. Env. Health 7: 76.

64. Smith, H. D. et al. (1963). The sequelae of pica with and without lead poisoning. Am. J. Dis. Child. 105: 609.

65. Chisolm, J. J., Jr. Unpublished data.

66. Sandstead, H. H., et al. (1969). Lead intoxication and the thyroid. Arch. Intern. Med. 123: 632.

67. Nishimura, H. (1964). Chemistry and Prevention of Congenital Anomalies. Charles C. Thomas, Springfield, Illinois, p. 13.

68. Gillet, J. A. (1955). Outbreak of lead poisoning in Canklow District of Rotterdam. Lancet 1: 1118.

69. Potter, E. L. (1961). Pathology of the Fetus and Newborn. (2nd ed.) Yearbook Medical Publishers, Chicago, p. 168.

70. Mao, P. and Molnar, J. J. (1967). The fine structure of lead-induced renal tumors in rats. Am. J. Path. 50: 571.

71. Van Esch, G. J. and Kroes, R. (1970). The induction of renal tumors in rats. Brit. J. Can. 23: 765.

72. Oyasci, R. (1970). Induction of cerebral gliomas in rats with dietary lead subacetate and 2-acetylaminofluorene. Cancer Res. 30: 1248.

73. Hammond, P. B. (1969). Lead poisoning, an old problem with a new dimension. In: F. R. Blood (ed.). Essays in Toxicology. Academic Press, New York, p. 115.

74. Goldwater, L. J. and Hoover, A. W. (1967). An international study of "Normal" levels of lead in blood and urine. Arch. Env. Health 15: 60.

75. Thomas, H. V. et al. (1967). Blood lead of persons living near freeways. Arch. Env. Health 15: 695.

76. Bradley, J. E. et al. (1956). The incidence of abnormal blood levels of lead in a metropolitan pediatric clinic. J. Pediat. 49:1.

77. Millar, J. A. (1970). Lead and δ-aminolevulinic acid dehydratase levels in mentally retarded children and in lead-poisoned suckling rats. Lancet 2: 695.

78. Hernberg, S. et al. (1970). δ-Aminolevulinic acid dehydrase as a measure of lead exposure. Arch. Env. Health 21: 140.

79. Bonsignore, D., Calissano, P., and Cartasegna, C. (1965). A simple method for determining δ-aminolevulinic dehydratase in the blood. Med. Lavoro 56: 199.

80. DeBruin, A. (1968). Effect of lead exposure on the level of δ-aminolevulinic dehydratase activity. Med. Lavoro 59: 411.

81. Nakao, K., Wada, O., and Yano, Y. (1968). δ-aminolevulinic acid dehydratase activity in erythrocytes for the evaluation of lead poisoning. Clin. Chim. Acta 19: 319.

82. Hernberg, S., et al. (1970). Erythrocyte ALA-dehydratase as a test of lead exposure. International Conf. on Chemical Pollution and Human Ecology, Prague.

83. Selander, S., and Cramer, K. (1970). Interrelationships between lead in blood, lead in urine, and ALA in urine during lead work. Brit. J. Ind. Med. 27: 28.

84. Feldman, F., et al. (1969). Serum δ-aminolevulinic acid in plumbism. J. Pediat. 74: 917.

85. Chisolm, J. J., Jr. (1968). The use of chelating agents in the treatment of acute and chronic lead intoxication in childhood. J. Pediat. 73: 1.

86. Hernberg, S., Nurminen, M. and Hasan, J. (1967). Non-random shortening of red cell survival times in men exposed to lead. Env. Res. 1: 247.

86. Selander, S., Cramer, K. and Hallberg, L. (1966). Studies in lead poisoning. Brit. J. Ind. Med. 23: 282.

88. Crutcher, J. C. (1963). Clinical manifestations and therapy of acute lead intoxication due to the ingestion of illicitly-distilled alcohol. Ann. Intern. Med. 59: 707.

89. Barltrop, D. (1967). The excretion of delta-aminolevulinic acid by children. Acta. Paediat. Scand. 56: 265.

90. Schwartz, S., Zieve, L. and Watson, C. J. (1951). An improved method for the determination of urinary coproporphyrin and an evaluation of factors influencing the analysis. J. Lab. Clin. Med. 37: 843.

91. Hsia, D. Y. Y. and Page, M. (1954). Coproporphyrin studies in children. 1. Urinary coproporphyrin excretion in normal children. Proc. Soc. Exp. Biol. Med. 85: 86.

92. Barltrop, D. and Killala, M. J. P. (1967). Fecal excretion of lead by children. Lancet 2: 1017.

CHAPTER 4. CADMIUM

D. W. FASSETT, Laboratory of Industrial Medicine,
Eastman Kodak Company, Rochester, New York

Cadmium was discovered by F. Strohmeyer, a metallurgist, in 1817, as a result of the study of some impurities found in zinc carbonate. The sulfide was noted to be a particularly beautiful yellow pigment and this remained the principal use of cadmium up until recent times (1). Cadmium is a transition metal in Group IIb along with zinc and mercury. The metal is silvery white in color, malleable, and ductile. The oxide is brown in color; the sulfide is a brilliant yellow, very insoluble pigment. Many salts, however, are quite soluble in water. Although the surface oxidizes readily, it is very resistant to rusting. It has a very valuable property of having a low melting point and forms several important low melting alloys. It forms hydroxides and complex ions with ammonia and cyanide; for example, $Cd(NH_3)_6^{4-}$ and $Cd(CN)_4^{2-}$. It also forms a variety of complex organic ammines, sulfur complexes and chelates such as those with acetonylacetone. There are eight stable isotopes known to be present in nature. It is worth noting that the lanthanide series intervenes between cadmium and mercury in the periodic table and that the electronic structure of zinc and cadmium differs fundamentally from that of mercury. The latter has an extra fourteen electrons in the 4th orbital which may account for its marked covalency and its tendency to form stable Hg-carbon bonds. Cadmium-carbon bonds can be formed but they are extremely unstable compared with mercury (2).

ECOLOGY

SOURCES

Geological

Cadmium is a relatively rare element, averaging one-half gram per ton of the earth's crust. It occurs as a mineral greenockite, but most of it is obtained from lead and zinc sulfide ores, particularly sphalerite, in the course of extracting those metals. A small amount is also recovered from flue dust in the refining or smelting of complex lead and copper ores. Table 1 gives common concentrations of cadmium and other metals in various rocks, but the content of pockets may rise considerably above these values. Drainage, microbiological activities such as those of sulfur bacteria, and pH may be of major importance in determining local concentrations (3-5).

Industrial

World production of cadmium has risen steadily since 1910 when most of it was produced in Germany and the level of production was only about 100,000 lb/yr. At the present time it is greater than 31,000,000 lb/yr (6). The United States produces about 40% of the world output. Next to the United States in production are the U.S.S.R., Canada, Japan, and Belgium. These five countries account for 81% of the total world's production. Mexico and Australia are also relatively large producers of cadmium. In the roasting or sintering of concentrates prior to smelting, cadmium is driven off as an oxide fume with lead and zinc oxide and is collected in baghouses or electrostatic precipitators. These recovery devices, of course, greatly reduce potential losses to the atmosphere.

In the electrolytic zinc process, roasted zinc concentrates or impure zinc oxide smelter dusts are dissolved in sulfuric acid. As one of the steps in purifying the resulting electrolyte, a cadmium metal sponge is precipitated by the addition of zinc dust. Final production of pure cadmium, whether from impure sponge or oxide-bearing dusts, is normally accomplished by electrolysis.

TABLE 1 -- OCCURRENCE OF CADMIUM IN ROCKS AND SOIL (DRY) (4)
ppm

Element	Igneous	Shales	Sandstones	Limestone	Coal	Soil
Cd	0.2	0.3	0.05	0.035	0.25	0.06
Cu	55	45	5	4	300	20
Hg	0.08	0.4	0.03	0.04	-	0.03
Pb	12	20	7	9	5	10
Zn	70	95	16	20	40	50

The relatively low melting point and high volatility of cadmium are extremely important in making practical its recovery from smelting processes. High volatility of course means a greater potential hazard in the production of cadmium and in its subsequent industrial uses.

Surveys have shown that there are a large number of consumers of cadmium in the United States. In a survey made in 1960, for example, there were about 1,526 respondents of whom 1400 gave electroplating as a principal use. Cadmium-plated parts find wide use in a variety of industries, including the automobile industry. Table 2 indicates the principal uses of cadmium, more or less in order of quantities used (1,7).

About 58% of the cadmium consuming industry in the United States is located in states which border the Great Lakes. The New England area also uses a considerable portion of the cadmium production. In addition to electroplating, which probably makes up some 60% of the total usage of cadmium, it finds wide uses in a variety of bearing alloys of particularly valuable properties. A variety of solders contain cadmium, e.g., silver solder. The pigments have extremely wide use in both paints and plastics since they are very stable and form extremely attractive yellow, orange, and reddish colors in a variety of tints. Cadmium yellow is essentially cadmium sulfide. Cadmium red consists of a mixture of cadmium sulfide and cadmium selenide with some added barium sulfate. Nickel-cadmium batteries are coming into wider use in a variety of ways. Semi-conductors and photocells form another, more recent type of application.

Organocadmium compounds such as dimethyl and diethyl cadmium are the principal ones used in

99

TABLE 2 -- PRINCIPAL USES OF CADMIUM

Electroplating (auto industry)
Pigments and Chemicals
Plastic stabilizer - Polyvinyl Cl
Alloys and Solder (silver solder and
    bearings)
Ni-Cd batteries
Semi-conductors - Photocells
Pesticides

Apparent Consumption in U. S. In
    1966 = 14,780,000 lbs.

commerce as polymerization catalysts. Diethyl cadmium has been used in tetraethyl lead synthesis (8). Some plastics may contain small quantities of cadmium salts stabilizers. Although there are about twelve pesticides known to contain cadmium, their total usage is probably relatively small and limited mainly to turf fungicides.

There are about thirty or more occupations where potential cadmium exposure may occur. Cadmium also occurs as a contaminant in superphosphate fertilizers in very small quantities. However, because of the large usage of such fertilizers, this has been the cause for some consideration of potential exposures from this source (1,7).

CYCLES AND DISTRIBUTION IN THE ENVIRONMENT

The pathways by which cadmium and its salts move through the environment and impinge upon man are very similar to those indicated in a previous chapter for lead, with the partial exception of auto emissions. The factors affecting the cycling of cadmium through the environment are not at all clear but it appears likely that the natural geochemical pollution is the major factor except for localized instances. (See later -- Itai-Itai Disease) The ability of plants to concentrate cadmium and their decomposition and dispersion in the environment may be an important factor.

The situation prevailing with cadmium is, in many respects, the opposite of that described for lead in a previous chapter. Man-made contributions to the environment are relatively recent and localized; most of the material found in soils,

TABLE 3 -- OCCURRENCE, RESIDENCE TIME AND PERCENT RETENTION
IN SEA WATER (4)

Element	ppm	Residence Time (yrs. x 10^3)	% Retention
Cd^{2+}	0.00011	500	0.05
Cu^{2+}	0.003	50	0.005
$HgCl$	0.00003	42	0.03
Pb^{2+}	0.00003	2	0.000002
Zn^{2+}	0.01	180	0.01

water, and food away from specially polluted areas
can be regarded as natural or background content.

Sulfides in the soil are generally somewhat
unstable materials and can be attacked by
microorganisms and made available for uptake by
plants. Plant cover can mobilize as much as 30 to 70
tons of minerals/km^2 (3). In general, the biological
cycles mobilize annually quantities several times
greater than the total that is transported to the sea
by rivers.

As can be seen from Table 3, lead has a
relatively short residence time in sea water compared
with cadmium and the total lead calculated to be
delivered to the ocean disappears relatively rapidly.
(4).

One of the most extensive studies of metal
content in various fresh waters is by Kopp (5). In
this study some 30,000 analyses were made on 1500 raw
and 380 finished waters in the United States. Table
4 gives the frequency with which four metals were
found and also the mean and range in $\mu g/li$. It can
be seen that cadmium is encountered very infrequently
in either raw water or more particularly in finished
water. The concentration levels for cadmium were
also considerably lower than those for copper, lead,
and zinc. Studies made on the leaching of certain
metals from a piece of copper piping containing a
soldered joint showed that, although copper, lead,
and zinc increased, no change was seen in cadmium.

In another study of community drinking water
supplies at the point of use, cadmium was below the
drinking water standard of 0.01 mg/li in over 99.8%
of 2,595 samples (see Table 3a). Lead, zinc, and
copper exceeded the standards somewhat more
frequently than cadmium (9).

TABLE 4 -- OCCURRENCE IN U.S. FRESH WATERS

A - (1500 raw and 380 finished waters) (5)

Element	Frequency % Raw	Frequency % Finished	Conc. μg/li Raw Mean	Conc. μg/li Raw Range	Conc. μg/li Finished Mean	Conc. μg/li Finished Range
Cd	2.5	0.2	9.5	(1-120)	12	-
Cu	75	65	15	(1-280)	43	(1-1056)
Pb	19	18	23	(2-140)	34	(3-139)
Zn	77	77	64	(2-1183)	79	(3-2010)

B - Frequency of occurrence and levels of cadmium in drinking water supplies (9)

Concentration mg/li	Frequency No. Samples
.000-.001	962
.002-.003	782
.004-.005	655
.006-.007	129
.008-.009	38
.010-.014	25
.015-.019	1
.020-.024	1
.025-.029	0
.030-	2
Total	2595

Bowen (4) has reviewed information on the occurrence of cadmium and a variety of other elements in plants and also in various species of animals. It is clear from the wide variety of plant and animal tissues examined over a number of years that cadmium content is of a very low order of magnitude. Perhaps the most striking exception is the finding of relatively high levels in mammalian liver and especially kidney. The apparent reason for this will be discussed later.

It is possible to calculate a concentration factor by dividing the parts per million in a fresh organism by the parts per million in water. Bowen (4) indicates that cadmium had a concentration factor of about 900 from sea water in the case of plankton and brown algae and a mean concentration factor of 1620 in the case of thirty-two fresh water plants. This is very considerably lower than the concentration factors for plankton in the case of copper, lead, and zinc and, also, for mercury and zinc in the case of fresh water plants. Nevertheless it is apparent that considerable concentration of cadmium can occur by certain water plants without apparent injury to the plant.

Hodgson (5) points out a number of factors that affect the movement of a mineral from a rock to a plant; e.g.: mechanical factors relating to the dispersion of the rock, oxidation-reduction properties, effects of biological organisms, crystallization, pH, etc. The nature of the ions or complexes that are present in soil and their relative solubility at various pH's affect the ability of plants to utilize these materials. Table 5 gives some data on concentrations and a plant/soil ratio is calculated. In this study cadmium appears to have a ratio considerably higher than lead, zinc, or copper, but the significance of this is uncertain. The results appear to vary from the concentration factors for fresh water plants given above. Cannon (5) also found that there might be considerable concentration of cadmium in plants under certain conditions in bogs studied in Western New York State. Whether these findings apply generally to the ability of plants to concentrate cadmium is unknown.

With the use of sensitive analytical methods such as atomic absorption, it can be shown that cadmium occurs in relatively small amounts in nearly all foods and beverages used by man (10) (Table 6). Fish and meats tend to show mean levels considerably higher than those for milk and eggs, cereals, and vegetables. Oysters and canned anchovies gave values of 3-5 ppm (wet weight). Some fats also tended to have higher content than vegetables, fruits, or nuts. The values quoted by Schroeder for various wheat flours seem to be higher than those given by Morris and Greene (11). These authors found values about one quarter or less of those given by Schroeder for

TABLE 5 -- RATIO OF SOIL TO PLANT Cd (5)
(ppm dry weight)

Element	Soil	Plants	Plant/Soil
Cd	0.06	0.64	10
Zn	50	32	0.6
Pb	10	4.5	0.45
Cu	20	9.3	0.45

TABLE 6 -- CADMIUM IN VARIOUS CLASSES OF FOODSTUFFS
(Modified from Schroeder et al. (10))

Class of food	Mean Concentration (ppm wet weight)
Seafood	0.79 (oysters excluded)
"	0.33 (oysters and anchovies excluded)
Meats	0.88
Dairy Products	0.27
Cereals and Grains	0.19
Vegetables (legumes)	0.03
Vegetables (tubers)	0.07
Vegetables (leafy)	0.13
Oils and Fats	0.83
Nuts	0.05
Fruits	0.04

flour, although baked bread showed somewhat higher values. The increase was considered to have come from sources other than flour. Morris and Greene also studied other dietary items in ten cities and concluded that variations seen were not due to a difference in wheat products. The values obtained by Murthy et al. (12), in repeated monthly sampling for the levels of cadmium and other trace metals in the total diets of institutionalized children (ages 9 to 12), in 28 U. S. cities, are considerably lower than those estimated by Schroeder for single food samples. The mean cadmium intake was 0.092 mg/day with a range of 0.032 to 0.158 mg/day. The mean value for zinc was 9.07 mg/day and the cadmium-zinc ratio was also lower than that given by Schroeder (0.006 compared to 0.015).

Duggan and Lipscomb (13) included cadmium in the so-called market basket study of 82 food items obtained in retail stores in 1968. The average daily intake was estimated to be 0.026 mg, about one-tenth of that estimated by Schroeder. Cadmium was present in all food classes examined and in about 12 percent of the composite samples.

TOXICOLOGY

PHARMACOLOGICAL ASPECTS

Information about the uptake, metabolism, and elimination of cadmium is derived mainly from experimental studies in which animals have been given the compounds orally or by injection, from individuals occupationally exposed, or from autopsy

104

TABLE 7 -- BALANCE STUDIES IN HUMANS WITH CADMIUM, ZINC, and COPPER (5)

mg/day Mean Values

Element	Subject	Diet	Feces	Urine	Feces & Urine	Diet minus Excreta	Urine Excreta
Cd	C	0.10	0.03	0.08	0.11	-0.010	0.75
	D	0.22	0.047	0.092	0.14	0.080	0.66
	E	0.18	0.049	0.11	0.16	0.02	0.69
Cu	C+D+E (Range)	.95-6.2	1.3-4.2	.022-0.042	1.3-4.2	-3.+2.0	.007-.032
Zn	C+D+E (Range)	11-18	14-16	0.55-1.3	15-17	-4.0-1.0	.034-.087

material (14). There have been very few carefully designed human balance studies. The available evidence is often conflicting, partly because of some doubts about the accuracy of methods used to measure small amounts.

Uptake and Elimination

There appear to be no special features about the deposition of cadmium in the respiratory tract from inhaled fumes. That which is deposited has a half-life of about 5 days. Most of the removal is by transport to other parts of the body (Friberg, personal communication). The sulfide tends to be trapped to a greater extent in the lungs than other forms (15).

In balance studies carried out by Tipton and Stewart (5) with very small amounts of cadmium (not more than .2mg/day) in the diet, the apparent retention of cadmium was extremely small. The substance did, however, appear to be absorbed and eliminated in the urine, rather than passed directly on in the feces (Table 7). A variety of other reports on the fate of cadmium salts injected into the animals suggests, however, that the major route of excretion is into the alimentary canal and thence through the feces (10). Still other reports (Friberg, personal communication) indicate that the amount of cadmium taken up by the body from that in the diet seldom exceeds 20% of a single dose, or 1% of a continued dose, the remainder being passed on and eliminated in the feces. It is probable that the normal value for urinary excretion is less than 5 ug/day. Imbus et al. (16) reported a mean value of 1.6 μg/ll for urinary excretion. Friberg believes that most values will be in the range of 1-2 µg/day in the urine.

Attempts have been made to investigate the comparative percutaneous absorption of metal compounds through guinea pig skin. Skog and Wahlberg (17) used a ^{115}Cd isotope on the skin of guinea pigs and determined the disappearance rate for a skin depot by use of a scintillation detector. Under these conditions cadmium was less well absorbed than three mercury compounds and sodium chromate. Only 1.8% of the total dose was absorbed in a period of five hours. It could be concluded that the cadmium chloride used in the study was not very readily absorbed percutaneously.

Transport, Storage, and Metabolism

Cadmium is transported in the blood from the site of absorption, but the concentration in "normal" blood is very low. In one study of 243 samples from males in 19 U. S. cities, the concentrations varied greatly among samples within a given location (18). The median concentration was about 0.5 µg/100ml, and about half of the samples showed no detectable amounts. The median value found by Imbus et al. (16) was 0.7 µg/100ml of whole blood.

Administration of soluble cadmium salts by mouth or by injection results in increased concentrations in the liver and kidneys.

As part of a study of chronic toxicity of cadmium administered in drinking water to rats, Decker et al. (19) determined the concentrations of cadmium in the liver and kidney. The levels in the drinking water varied from 0.1 ppm to 50 ppm.

The rats were followed over a period of 12 months. Both the kidney and the liver tended to show increased cadmium values more or less in proportion to the dose. Some rats were also sacrificed after six months' feeding and results showed that there was an increase in storage in both liver and kidney proportional to time. Approximately 0.3 to 0.5% of the total dose of cadmium ingested in one year was retained by the kidneys and liver. As others have found, the concentration of cadmium in kidney tissue was generally several times that in the liver.

However, the total amount of cadmium found in the liver at the end of the twelve-month period was about three times that found in the kidney. There was no evidence of any significant storage in the bone. No pathological findings were noted in the groups receiving between 0.1 and 10 ppm. At 50 ppm of cadmium in the drinking water for a period of three months there was a definite reduction in growth rate, water consumption, and blood hemoglobin. Bleaching of the incisors was also noted.

In an extension of the above study, Anwar et al. (20) studied nine female dogs over a period of four years, during which time their drinking water contained cadmium concentrations varying from 0.5 to 10 ppm. At the end of the study these dogs were sacrificed and samples of kidney, liver, and pancreas were prepared and analyzed for cadmium with the same method as in their rat study. Histological studies were also made. As in the case of rats, cadmium accumulated in liver and kidney tissues of dogs in amounts proportional to the concentration of the element in drinking water up to 5 ppm. There was little difference in cadmium liver and kidney concentrations between the 5 and 10 ppm levels. The pancreas showed little accumulation of cadmium at any level. There were no changes in urine or blood and the only significant histologic change of uncertain significance was a slight increase in fat in the proximal convoluted tubules of the kidney.

Valuable information about the storage of cadmium in the body following occupational exposures has been developed over the past fifty years. One of the earliest reports was by Stephens (21) and described the case of a 67-year-old man who had worked in a zinc smelting operation for many years. Although his diagnosis was said to be that of lead poisoning, Stephens noted that the symptoms did not match those generally associated with lead exposures. A post-mortem examination was done and it was found that the liver contained no lead and only a trace of copper. However, cadmium was found in a quantity of about 120 mg/kg of liver. Zinc was also present at a level of 102 mg/kg. Eight other cases were eventually found where increased cadmium content was noted in the liver at autopsy. Other more recent examples can be found in the descriptions by Kazantzis et al. (22), Bonnell (23), Friberg (24), and Smith et al. (25). In all cases the liver and kidneys show relatively high concentrations compared to those found in normal unexposed subjects. In some cases the liver concentrations are higher than those of the kidney. This may of course be related to the nature of the cadmium compound inhaled. In the reference by Kazantzis the patient in question had been exposed principally to cadmium pigments, presumably cadmium sulfide or selenide, although he had also worked on other types of cadmium compounds.

The explanation for the remarkable retention of cadmium in the liver and kidney of many species is undoubtedly related to the discovery by Margoshes and Vallee (26) in 1957 that equine renal cortex contains an unusual protein which they have named metallothionein. This protein has been demonstrated to be present in the human renal cortex and in nature it is very similar to that found in the equine renal cortex (27). The equine form of metallothionein was shown to contain as high as 5.9% cadmium and 2.2% zinc. The metal-free protein, thionein, contained a high percentage of sulfur, 9.3%. Human metallothionein obtained thus far appeared to contain very similar quantities of cadmium, 4.2%, and also zinc, 2.6%. Cadmium is more firmly bound than zinc in this protein and presumably these two elements can compete for binding sites.

It has recently been shown by Shaikh and Lucis (28) that the formation of this cadmium binding protein can be induced in male and female rats by both subcutaneous injection and by oral administration in the drinking water in low concentrations (0.5 mM/li). The authors also studied the formation of this protein following doses of a mixture of ^{109}Cd and ^{65}Zn. They suggest that cadmium is probably poorly absorbed but that the absorbed fraction does induce the synthesis of cadmium binding from protein. The studies also showed that ^{65}Zn content of the liver and kidneys declined rapidly three weeks after exposure whereas ^{109}Cd showed only minor changes.

It seems possible that the induction of metallothionein might have been responsible for the finding by Terhaar et al. (29) that testicular atrophy and death from orally administered cadmium chloride can be prevented by pretreatment with small oral doses of cadmium chloride. Doses of cadmium chloride as low as 10 μg/kg given orally 24 hours prior to a dose of 100 mg/kg completely protected against massive testicular atrophy. Whether or not the induction of this protein by pretreatment with an ion such as cadmium or zinc would account for some of the antagonistic effects of zinc against cadmium is unknown but seems plausible.

Site of Action

Cadmium has been studied for its effects on various enzyme systems (14,30,31). It would appear as though cadmium is able to inactivate -SH containing enzymes in vitro and that it can produce an uncoupling of oxidative phosphorylation in very low concentrations in mitochondria. On the other hand, authors have also noted protective effects of a prior exposure to cadmium. For example, Gunn, et al. (32) have shown that one year after testicular damage by cadmium exposure regenerative vessels on the testis would show no evidence of injury following a second injection of cadmium. Gabbiani et al. (33) have shown that cadmium injected parenterally in rats induces hemorrhagic lesions in sensory ganglia. These lesions heal within about two weeks after which the animals become tolerant to a second injection of cadmium. Yoshikawa (34) has also noted that the pretreatment of mice with a small dose of cadmium chloride (0.6 mg/kg) 24 hours prior to a dose of 3 mg/kg of cadmium prevented the testicular damage, as reported earlier by Terhaar et al. The induction of the cadmium binding protein by pretreatment with a small dose would seem to offer an attractive explanation for these effects. Piotrowski (personal communication) and his co-workers (35) have also produced evidence for the induction of metallothionein and regard this as having an important role in preventing non-specific binding to SH- enzymes in vivo, such as appears to be the case with mercury.

The mechanism of protection by zinc against cadmium induced toxic effects such as those on the testis has been studied intensively by Gunn and his coworkers (36). It was first established by Parizek (37) that zinc could exert an extraordinary

109

protective effect against the acute atrophy of the testis from exposure of cadmium. Gunn et al., studied the protective effects of zinc, cystein, and selenium against cadmium testicular atrophy. Using [109]Cd it could be shown that none of these protective agents actually decreased the amount of cadmium reaching the testis. However, selenium did induce increased intake of cadmium in the testis. It is thought that the primary action of cadmium testicular damage results from vascular effects (38). The conclusion was drawn that these protective agents probably are exerting their effects at the vascular level. Whether induction of metallothionein would play a role in such protection is unknown.

Friberg (39) described the occurrence of an unusual protein in the urine of persons chronically exposed to cadmium by inhalation. The nature and origin of this protein have been the subject of numerous publications (40-42).

Piscator (42) has shown that workers with chronic exposure to cadmium excrete considerable quantities of a series of proteins of low molecular weight (10,000-30,000) but that these proteins actually are not unusual in their structure, consisting of a small albumin fraction and large alpha 2, beta and gamma fractions. Vigliani (43) has proposed that cadmium may interfere in some way with the manufacture or disposal of immunoglobulins since there is some evidence that the low molecular weight proteinuria caused by cadmium might be related to the l-chain of immunoglobulins. A suggestion has been made that perhaps cadmium does play a physiological role in regulating the biosynthesis of molecular variations of albumin and perhaps other proteins (44).

Mild hypochromic anemias are common in cadmium exposures and this has been reproduced in rabbits (45). The exact mechanism of cadmium anemia has not been completely worked out. Friberg noted that rabbits treated with cadmium excreted considerably increased quantities of iron in the urine. Feeding of iron had some prophylactic benefit but liver had no effect on the anemia. Friberg suggested that the excreted iron was in all probability bound to a protein, although no experiments were carried out to demonstrate this. No reports of experiments to study the possibility that cadmium might displace iron from transferrin could be found in the literature. A yellowish discoloration of the teeth has occasionally been noted. Possibly this could be due to the formation of a sulfide. Japanese quail given a diet containing 75 mg/kg of cadmium could be protected against anemia and poor growth by large amounts of ascorbic acid (10,000 mg/kg of diet). d-iso-ascorbic acid was also effective. A later report showed that the spermatogenic effect could also be prevented in this species by a large amount of ascorbic acid (46,47).

CLINICAL ASPECTS

Health Effects

Cadmium has been known since 1858 to exert toxic effects. Knowledge of the effects on man has come primarily from many studies that have been made on occupationally exposed people as well as from accidental poisoning by ingestion and by abortive attempts to use it therapeutically. There are a number of excellent reviews of the history of occupational cadmium poisoning: Stokinger (14), Prodan (48), Bonnell (49), Kazantzis et al. (22), Kazantzis (50), and Friberg (39). Table 8 lists some of the exposure levels at which cadmium is known to produce toxic effects in man and other species.

The acute effects of oral ingestion in man are those of immediate nausea and vomiting which can occur from as little as 15 mg of total cadmium. There has been a considerable number of epidemics of acute nausea following the attempts to use cadmium plated articles as food containers (51,52).

Both acute and chronic effects have been noted as a result of occupational exposure. Acute effects by inhalation come from exposure to cadmium oxide fume or to high concentrations of dust. A typical situation would be the cutting or welding of a cadmium containing object and the symptoms would be those of somewhat delayed pneumonitis, including coughing, pain in the chest, and finally a severe chemical pneumonitis. Fatalities have occurred from exposure to cadmium fume at levels from 3-100 mg/m^3

TABLE 8 -- SOME TOXIC LEVELS OF CADMIUM EXPOSURE

Species	Route	Concentration	Time	Effects
Man	Oral (soluble salts)	>15 mg	Single dose	Vomiting
Man	Inhalation (CdO fume)	0.5-2.5 mg/m³	Intermittent 2-3 days	Pneumonitis
Man	Inhalation (occupational exposure) (CdO fume and dust)	Unknown but >0.2 mg/m³	Chronic (many years)	Emphysema Mild anemia Proteinuria
Rats	Oral (drinking water)	10-50 ppm	1 year	Anemia and reduced growth rate
Rats	Oral (diet)	135 ppm	6 months	Anemia and poor growth
Dogs	Oral (drinking water)	5-10 ppm	4 years	No significant toxic effects

(53).   Non-fatal acute pneumonitis has been observed during silver soldering from intermittent exposure to cadmium fume concentrations of from 0.5-2.5 mg/m³ over a 3-day period (54).   Recovery may take as long as two weeks or more.

Chronic exposure produces a rather characteristic emphysema which is often unaccompanied by previous history of chronic bronchitis or coughing. Such emphysema may be extremely disabling and may progress even after exposure has ceased.

An altered pattern of urinary excretion of amino acids has been observed from exposure to cadmium by Clarkson and Kench (55). In particular, there was a marked increase in the hydroxy amino acids threonine and serine. Clinically, the disturbance in renal function somewhat resembles that found in conditions associated with renal tubular malfunction; for example: renal glycosuria, aminoaciduria, proteinuria, impaired concentrating ability, impaired acid excretion, etc. In general, the degree of disability from the renal tubular damage appears to be minimal.

Another symptom that has been mentioned is loss of smell (anosmia) (56). Other metal exposures (nickel) may have coexisted in some cases.

Because of the well-known effects of cadmium on the kidney, it was natural that clinical investigators should explore the possibility that hypertensive disease might be related to cadmium in the environment. In particular, Schroeder and his coworkers have discussed possible relationship in many publications (57). Perry has reviewed the basis for the theory that cadmium and human hypertension are related (58). The strongest argument in favor of this theory would appear to be based on the claim that humans with essential hypertension can be shown to have more cadmium in their kidneys or an altered Cd/Zn ratio compared to those without hypertension. Experimental animals such as rats and rabbits also are said to develop hypertension when exposed to cadmium. For example, Thind, et al. (59) produced hypertension in rabbits by a series of weekly intraperitoneal injections of cadmium acetate in doses of 2 mg/kg.

In spite of the attractiveness of the hypothesis that cadmium is responsible for human hypertension, there are reasons to question whether this theory is correct. For example, Morgan (60) studied samples of kidney and liver tissue obtained postmortem from 80 individuals having various types of cardiovascular disease, including a hypertension group, and also some with neoplastic disease and some who had none of these diseases. She found no significant difference between the control group and those with either hypertensive or other forms of cardiovascular disease. The neoplastic disease group, however, did appear to be from a somewhat different population distribution.

Szadkowski et al. (61) measured cadmium excretion in the urine of 169 persons not occupationally exposed to cadmium and studied the relation of cadmium excretion to age. No significant correlation could be found between the arterial pressure and cadmium excretion, although there was a slight but statistically valid increase in cadmium excretion (expressed as $\mu g/g$ of creatinine) with age.

Industrial physicians have also disputed the claim that cadmium is related to hypertension since workers occupationally exposed over long periods show little or no evidence of having an unusual incidence of hypertensive disease (62,63). It would seem best to conclude, as does Perry (58), that the significance of cadmium in relation to hypertension will have to wait for additional information.

113

Some attention has been given to the possibility that cadmium in the environmental air might be related to cardiovascular disease (64). Although some statistical correlations seem to exist, Carroll (5) states recently that further studies have not shown the correlation to be nearly as strong as in the original report. Carroll (5) also considers the relationship between inhaled cadmium from the environment and a total intake of cadmium. He points out that the average dietary intake of cadmium in the United States is probably in the order of 100 - 200 μg/day. (This assumption is probably in error as the mean intake appears lower.) The concentrations of suspended cadmium in urban air have a wide variation but probably average only 0.05 μg/m^3. Assuming an air intake of about 15 m^3/day, dosage by inhalation would only amount to 0.75 μg/day or less than 1% of the dietary intake. He suggests that, since recent measurements of cadmium in dustfall in urban areas indicated that over 2 grams of cadmium might be deposited on an acre per year, that some attention should be paid to deposition of cadmium from the atmosphere as a possible indirect source of increasing the dietary intake. A recent epidemiologic study in 77 midwestern cities showed, however, that there was no apparent correlation between the cadmium content of dustfall and cardiovascular death rates in urban areas. There was also no association between cadmium concentration in milk and cardiovascular disease (65).

Because of the well-known effects of cadmium in the testis of experimental animals, as discussed previously, a number of investigations have been made as to the effect of cadmium on the course of pregnancy and on fertility. Parizek (66,67) injected cadmium salts subcutaneously in pregnant rats between the 17th and 21st day of gestation. The dose used was 0.04 mMol/kg body weight. This caused very rapid destruction of the fetal placenta with only very slight effects on the maternal placenta.

In contrast to this, animals injected with a combination of cadmium salts and sodium selenite showed no typical placental effects. The hemorrhagic renal necrosis which also accompanied cadmium intoxication during the last four days of pregnancy was prevented by the addition of sodium selenite.

The effect on male rats receiving a relatively low dose of cadmium chloride subcutaneously (1.25 mg/kg body weight) was studied by Kar et al. (68). The treated males were then mated with untreated females and the effects on fertility studied. In spite of the pronounced effects on the morphology of the testis, the effects on fertility were negligible, indicating that the sperm were relatively unaffected by cadmium.

Several studies have been made to evaluate the carcinogenic effect of cadmium. Gunn et al. (69) have produced sarcomas in rats by injection subcutaneously, subperiosteally, and intramuscularly. No tumors were formed in ectodermal, endodermal, or epithelial mesodermal sites. In contrast to these results, a number of studies reviewed by Shubik and Hartwell disclosed very little evidence that cadmium was carcinogenic (70).

Whether cadmium plays any role in human carcinogenesis is very doubtful at this time. Potts (63) noted the cause of death in eight men out of a group of 74 who had had more than ten years' exposure to cadmium. Three had carcinoma of the prostate, one carcinoma of the bronchus, and one carcinomatosis. However, other authors studying the epidemiology of cadmium in occupational diseases do not appear to have been impressed with any unusual incidence of cancer, in spite of the obvious chronic effects of cadmium exposure on the lung and on the kidney. Butt (71) reported a study of cancerous tissues for a content of 13 trace metals and found that the average cadmium content was higher than those in control tissues. However, as pointed out by Morgan (60), malignant tissues do appear to have very erratic distributions of metals in general so that the significance of this is not known as regards the malignant process.

Since cadmium accumulates to a very high degree in the liver and kidney one might expect that these organs would be the site of any possible malignant changes; however, there appear to have been no epidemiologic studies indicating that this is occurring. The significance for human cancer of animal experiments in which relatively large quantities of reactive metals are injected in a

variety of sites is difficult to interpret with
reference to the case of very low level exposures in
the human environment.

Itai-Itai Disease

Considerable interest has been sparked recently
by reports of a syndrome occurring in Japan and
attributed by many to environmental pollution with
cadmium (72,73). Itai-itai (or ouch-ouch in English
onomatopeia) appears to be an endemic condition, seen
particularly in elderly females, characterized by
pain in the bones and joints, waddling gait,
aminoaciduria, glycosuria, severe osteomalacia, and
multiple pathological fractures (74). Cases began to
be observed about 1912 in an area potentially
contaminated with metals from a lead mine. The
number of cases increased about the time that the
mine began to produce zinc and cadmium. Over a
period of 15 years, which included the war years of
1939-45, some 200 persons living along the banks of
the Jintsu River suffered from the syndrome. Half of
them died. Women 50-60 years of age who had borne
many children were especially at risk.

There is considerable disagreement on the
etiology. Those who favor metallic contamination as
a cause, contend that the air, soil, run-off water,
and rice crops were successively polluted, and
produce sampling evidence in support. On the other
hand, some of the specific disturbances, and
particularly the osteomalacia, are not seen in other
undoubted cases of cadmium intoxication. Attempts to
reproduce the condition experimentally in animals
have not succeeded. Since new cases have not
appeared since 1955 it is difficult to correlate
cadmium intake from foods with the disease although
it is probable that the intake was elevated. The
affected areas undoubtedly were those with low
calcium and vitamin D intake and the presence of
other elements such as lead or fluoride may have been
involved. Tsuchiya and Takeuchi (personal
communication) believe that increased cadmium intake
alone did not produce the disease but that other
factors, e.g., nutritional, were essential.

At one end of the scale, a condition other than
cadmium intoxication (war-time malnutrition has been

cited) may be the essential agent. The possibility that cadmium pollution may, under suitable circumstances, produce the disturbances described should not be overlooked; but neither should the Japanese experience be used to suggest that extensive cadmium intoxication lies just around the corner. Much more evidence is necessary.

REFERENCES

1. Schroeder, H. J. (1965). Mineral Facts and Problems. Bureau of Mines Bulletin No. 630.

2. Sidgwick, N. V. (1950). The Chemical Elements and Their Compounds. Oxford University Press, New York.

3. Lukashev, K. I. (1958). Lithology and Geochemistry of the Weathering Crust. Izdatel'stvo Akademii Belorusskoi SSR, Minsk. Translated from Russian, Israel Program for Scientific Translations, Ltd. (1970).

4. Bowen, H. J. M. (1966). Trace Elements in Biochemistry. Academic Press, New York.

5. Hemphill, D. D. (Ed.) (1969). Trace Substances in Environmental Health, III. University of Missouri, Columbia, Missouri.

6. U. S. Department Interior (1968). Minerals Yearbook. U. S. Govt. Print. Off., Washington, D. C.

7. National Air Pollution Control Administration. (1969). Preliminary Air Pollution Survey of Cadmium and Its Compounds. NAPCA APTD-69-32. U. S. Govt. Printing Office, Washington, D. C.

8. Harwood, J. H. (1963). Industrial Applications of the Organometallic Compounds. Reinhold Publishing Corporation, New York.

9. Bureau of Water Hygiene. (1970). Community Water Supply Study. Environmental Health Service, Cincinnati, Ohio.

10. Schroeder, H. A. et al. (1967). Essential trace metals in man. Zinc. Relation to environmental cadmium. J. Chron. Dis. 20: 179.

11. Morris, E. R. and Greene, F. E. (1970). Distribution of lead, tin, cadmium, chromium and selenium in wheat and wheat products. Fed. Proc. 29: 500.

12. Murthy, G. K., Rhea, U. and Peeler, J. T. (1971). Levels of antimony, cadmium, chromium, cobalt, manganese and zinc in institutional total diets. Env. Sci. Tech., 5: 436.

13. Duggan, R. E. and Lipscomb, G. Q. (1969). Dietary intake of pesticide chemicals in the United States (II), June 1966 - April 1968. Pesticide Monitoring Journal 2: 153.

14. Stokinger, H. E. (1963). The metals (excluding lead). In: F. A. Patty, D. W. Fassett and D. D. Irish (eds.). Industrial Hygiene and Toxicology, Vol. 2. Wiley, New York.

15. Prodan, L. (1932). Cadmium poisoning: 2. Experimental cadmium poisoning. J. Ind. Hyg. Toxicol. 14: 174.

16. Imbus, H. R. et al. (1963). Boron, cadmium, chromium and nickel in blood and urine. Arch. of Env. Health 6: 286.

17. Skog, E. and Wahlberg, J. E. (1964). A comparative investigation of the percutaneous absorption of metal compounds in the guinea pig by means of the radioactive isotopes: Cr, Co, Zn, Ag, Cd, Hg. J. Invest. Derm. 43: 187.

18. Kubota, J., Lazar, V. A. and Losee, F. (1968). Copper, zinc, cadmium and lead in human blood from 19 locations in the United States. Arch. Env. Health 16: 788.

19. Decker, L. E. et al. (1958). Chronic toxicity studies. 1. Cadmium administered in

drinking water to rats. Arch. Ind. Health 18: 228.

20. Anwar, R. A. et al. (1961). Chronic toxicity studies. 3. Chronic toxicity of cadmium and chromium in dogs. Arch. Env. Health 3: 456.

21. Stephens, G. A. (1920). Cadmium poisoning. J. Ind. Hyg. 2: 129.

22. Kazantzis, G. et al. (1963). Renal tubular malfunction and pulmonary emphysema in cadmium pigment workers. Quart. J. Med., New Series 32, No. 126: 165.

23. Bonnell, J. A. (1955). Emphysema and proteinura in men casting copper-cadmium alloys. Brit. J. Ind. Med. 12: 181.

24. Friberg, L. (1957). Deposition and distribution of cadmium in man in chronic poisoning. Arch. Ind. Health 16: 27.

25. Smith, J. P., Smith, J. C. and McCall, A. J. (1960). Chronic poisoning from cadmium fume. J. Path. Bact. 80: 287.

26. Margoshes, M. and Vallee, B. L. (1957). A cadmium protein from equine kidney cortex. J. Am. Chem. Soc. 79: 4813.

27. Pulido, P., Kagi, J. H. R. and Vallee, B. L. (1966). Isolation and some properties of human metallothionein. Biochemistry 5: 1768.

28. Shaikh, Z. A. and Lucis, O. J. (1970). Induction of cadmium-binding protein. Fed. Proc. 29: 298.

29. Terhaar, C. J. et al. (1965). Protective effects of low doses of cadmium chloride against subsequent high oral doses in the rat. Toxic. Appl. Pharmacol. 7: 500.

30. Simon, F. P., Potts, A. M. and Guarde, R. W. (1947). Action of cadmium and thiols on tissues and enzymes. Arch. Biochem. 12: 283.

31. Jacobs, E. E. et al. (1956). Uncoupling of oxidation phosphorylation by cadmium ion. J. Biol. Chem. 223: 147.

32. Gunn, S. A., Gould, T. C. and Anderson, W. A. D. (1966). Loss of selective injurious vascular response to cadmium in regenerated blood vessels of testis. Am. J. Path. 48: 959.

33. Gabbiani, G., Baic, D. and Deziel, C. (1967). Studies on tolerance and ionic antagonism for cadmium or mercury. Canad. J. Physiol. Pharmacol. 45: 443.

34. Yoshikawa, H. (1969). Preventive effects of pretreatment with small doses of metals upon acute metal toxicity. 4. Cadmium. Igaku to Seibutsugaku 78: 211.

35. Wisniewska-Knypl, J. M. and Jablonska, J. (1970). Selective binding of cadmium in vivo on metallothionein in rat's liver. Bull. Acad. Polon. Sc. Cl 2. Ser. Sci. Biol. 18: 321.

36. Gunn, S. A., Gould, T. C. and Anderson, W. A. D. (1968). Mechanisms of zinc, cysteine and selenium protection against cadmium-induced vascular injury to mouse testis. J. Reprod. Fert. 15: 65.

37. Parizek, J. (1957). The destructive effect of cadmium ion on testicular tissue and its prevention by zinc. J. Endocrinol. 15: 56.

38. Mason, K. E. et al. (1964). Cadmium-induced injury of the rat testis. Anat. Rec. 149: 135.

39. Friberg, L. (1950). Health hazards in the manufacture of alkaline accumulators, with special reference to chronic cadmium poisoning. Acta Med. Scand. 138: Sup. 240.

40. Piscator, M. (1962). Proteinuria in chronic cadmium poisoning. 1. An electrophoretic and chemical study of urinary and serum proteins from workers with chronic cadmium poisoning. Arch. Env. Health 4: 607.

41. Smith, J. C., Wells, A. R. and Kench, J. F. (1961). Observations on the urinary protein of men exposed to cadmium dust and fume. Brit. J. Ind. Med. 18: 70.

42. Piscator, M. (1966). Proteinuria in chronic cadmium poisoning. 3. Electrophoretic and immunoelectrophoretic studies on urinary proteins from cadmium workers, with special reference to the excretion of low molecular weight proteins. Arch. Env. Health 12: 335.

43. Vigliani, E. C. (1969). The biopathology of cadmium. Am. Ind. Hyg. Assoc. J. 30: 329.

44. Editorial. (1968). Cadmium and metabolism of albumin. Lancet I: 133.

45. Friberg, L. (1955). Iron and liver administration in chronic cadmium poisoning and studies of distribution and excretion of cadmium; experimental investigations in rabbits. Acta Pharmacol. Toxic. 11: 168.

46. Fox, M. R. S. and Fry, B. F., Jr. (1970). Cadmium toxicity decreased by dietary ascorbic acid supplements. Science 169: 989.

47. Richardson, M. E. and Fox, M. R. S. (1970). Dietary ascorbic acid protection of cadmium-inhibited spermatogenesis. Fed. Proc. 29: 298.

48. Prodan, L. (1132). Cadmium poisoning: History of cadmium poisoning and use of cadmium. J. Ind. Hyg. 14: 132.

49. Bonnell, J. A. (1965). Cadmium poisoning. Ann. Occup. Hyg. 8: 45.

50. Kazantzis, G. (1956). Respiratory function in men casting cadmium alloys. Part 1. Assessment of ventilatory function. Brit. J. Ind. Med. 13: 30.

51. Fairhall, L. T. (1957). Industrial Toxicology. (2nd ed.) Williams & Wilkins Company, Baltimore, Maryland.

52. California State Water Pollution Control Board. (1957). Water Quality Criteria. (2nd ed.) Sacramento, California.

53. American Conference of Governmental Industrial Hygienists. (1966). Documentation of Threshold Limit Values. (Revised ed.) Cincinnati. Ohio.

54. Hygienic Guides Committee. (1962). Cadmium. Am. Ind. Hyg. Assoc. J. 23: 518.

55. Clarkson, T. W. and Kench, J. F. (1956). Urinary excretion of amino acids by men absorbing heavy metals. Biochem. J. 62: 361.

56. Adams, R. G. and Crabtree, N. (1961). Anosmia in alkaline battery workers. Brit. J. Ind. Med. 18: 216.

57. Schroeder, H. A. (1965). Cadmium as a factor in hypertension. J. Chron. Dis. 18: 647.

58. Hemphill, D. D. (ed.). (1968). Trace Substances in Environmental Health, II. University of Missouri, Columbia, Missouri.

59. Thind, G. S. et al. (1970). Vascular reactivity and mechanical properties of normal and cadmium-hypertensive rabbits. J. Lab. Clin. Med. 76: 560.

60. Morgan, J. (1969). Tissue cadmium concentrations in man. Arch. Intern. Med. 123: 405.

61. Szadkowski, Von D., Schaller, K. H. and Lehnert, G. (1969). Renal cadmiumausscheidung, lebensalter und arterieller blutdruck. Z. Klinische Chemie 7: 551.

62. Holden, H. (1969). Cadmium toxicology. Lancet 2: 57.

63. Potts, C. L. (1965). Cadmium proteinuria -- the health of battery workers exposed to cadmium oxide dust. Ann. Occup. Hyg. 8: 55.

64. Carroll, R. E. (1966). The relationship of cadmium in the air to cardiovascular disease death rates. J. Am. Med. Assoc. 198: 267.

65. Hunt, W. F., Jr. et al. (1970). A study of trace element pollution of air in 77 midwestern cities. Trace Substances in Environmental Health, 4th Annual Conference, University of Missouri.

66. Parizek. J. (1964). Vascular changes at sites of oestrogen biosynthesis produced by parenteral injection of cadmium salts: The destruction of placenta by cadmium salts. J. Reprod. Fert. 7: 263.

67. Parizek, J. et al. (1968). Pregnancy and trace elements: The protective effect of compounds of an essential trace element -- selenium --against the peculiar toxic effects of cadmium during pregnancy. J. Reprod. Fert. 16: 507.

68. Kar, A. B., Dasgupta, P. R. and Das, R. P. (1961). Effect of a low dose of cadmium chloride on the genital organs and fertility of male rats. J. Sci. Ind. Res. 20C: 322.

69. Gunn, S. A., Gould, T. C. and Anderson, W. A. D. (1967). Specific response of mesenchymal tissue to cancerigenesis by cadmium. Arch. Path. 83: 493.

70. Shubik, P. and Hartwell, J. L. (1951). In: J. A. Peters, (ed.) Survey of Compounds Which Have Been Tested for Carcinogenic Activity. PHS Publication No. 149. Sup. 1 (1957)., Supp. 2 (1969).

71. Butt, E. M. (1960). Trace metals in health and disease. Air Pollution Med. Res. Conf., San Francisco, California.

72. Kobayashi, J. (1970). Relation between "Itai-Itai" disease and pollution of river water by cadmium from a mine. Fifth International Water Pollution Res. Conf., San Francisco, California.

73. Tsuchiya, K. (1969). Cause of ouch-ouch disease (Itai-Itai Byo), an introductory review. Klio J. Med. 18: 181.

74. Editorial. (1971). Cadmium pollution and Itai-Itai Disease. Lancet I: 382.

PART II

OTHER CONTAMINANTS

CHAPTER 5. BERYLLIUM

LLOYD B. TEPPER, Kettering Laboratory, University
of Cincinnati College of Medicine, Cincinnati, Ohio

Beryllium in its pure form is a light, silvery metal similar in appearance to aluminum. Massive pieces of metallic beryllium are of no biological importance. When the metal or its compounds are present as airborne particulates of respirable size, however, a hazardous situation may exist since the inhalation of such materials has been associated with both acute and chronic illness. Beryllium dissolved in water or in food supplies is not known to be of health significance unless very high experimental concentrations are achieved. For all practical purposes then, the problem is one related to the concentration and size of beryllium particulates in industrial and ambient atmospheres.

ECOLOGY

SOURCES

Beryllium is not widely distributed in nature. The most important present source is beryl, a beryllium aluminum silicate mineral found in Brazil, Argentina, India, South Africa, and Rhodesia. In the United States, beryl is found in Colorado, Utah, South Dakota, New Mexico, and the New England States. It is of interest that pure beryl of gemstone grade forms the emerald. Beryl crystals are handsorted from ore materials to provide the primary raw material for metal production. As far as is known, beryl crystals and their dusts, because of physical, chemical, or epidemiological factors, have not been associated with human or experimental disease.

In the past several years, a beryllium mineral known as bertrandite has been mined on the Colorado plateau. This mineral contains soluble beryllium constituents which have been shown to be toxic in experimental animals. Hygienic considerations applicable to miners of these ores are based upon conventional industrial criteria.

Since beryllium is a chemically active metal, with a strong tendency to form compounds with other elements, the extraction of the metal from its ores and the isolation of the element in its pure form require relatively vigorous metallurgical processes. This primary smelting is performed by a small number of industrial firms in this country and abroad. Approximately 95% of the beryllium handled in domestic commerce is produced in northern Ohio or in the Reading-Hazelton area of Pennsylvania. The producers provide a broad spectrum of beryllium alloys and compounds and certain finished products. In addition they sell a "master-alloy" of relatively high beryllium content which may be used by purchasers to introduce beryllium into various alloys of lower beryllium content. Low beryllium alloys are distributed rather widely, either by the primary producers or by secondary vendors of metals and alloys.

USES

The great majority of cases of beryllium disease in this country has been associated with use of the metal which are now obsolete, viz., the production and use of beryllium-containing phosphors (fluorescent powers) for fluorescent lamps and "neon" signs. This production occurred almost entirely in the 1940's and resulted in several foci of endemic beryllium disease, the most notable of which is centered in Salem, Massachusetts. Cases from the "neon" lamp industry are more widely disseminated since this is a less formal industry with many operators working out of garages and small workshops. Industrial hygiene under such circumstances is often essentially nil.

Several contemporary applications utilize most of the current beryllium production. Over the past several years the use of metallic beryllium dust in combination with suitable oxidizers in solid-fueled rocket motors has demonstrated the technological advantage of such systems. Only the nuclear engine-equipped rocket has greater power in proportion to size than beryllium-burning rocket motors. Because of the apparent space and military advantages of high-power rockets, the examination and use of this system has been promoted by those in the relevant industries and military services. Since the fumes given off by burning beryllium are toxic, however, research and development along these lines has been inhibited by considerations of air contamination and the public health. Tests in which all effluent cannot be contained require control over large tracts of restricted lands. Even when tests are conducted under carefully prescribed meteorological conditions, limited-access areas may encompass zones of many square miles. For that reason the future of the beryllium rocket is uncertain.

Beryllium and its alloys are widely used in special structural applications in which specific properties are desired. Beryllium-containing bronzes are highly resistant to corrosion. Beryllium-copper can be used to make tools which do not cause sparks when struck against concrete or metallic surfaces. Beryllium itself is lighter, stronger, and more rigid than aluminum, and it therefore has obvious applications in the aerospace industry. Other beryllium alloys are used in spring metals.

Except in unusual situations, the general public is not at risk. Beryllium is not disseminated in ambient community atmospheres. Exceptions to this statement may be related to the discharge of beryllium dust from poorly controlled industrial operations. While hazardous operations of this type have been observed in the past, such conditions are not properly associated with modern industrial practice. Small amounts of beryllium, probably contained in beryl, may be detected in the air of communities using coal as fuel. This beryllium content of coal and exhausts is of no known hygienic importance. Such beryllium may be deposited in small amounts in the lungs and may be measurable on assay. Beryllium as it presently occurs in communities does not represent a factor in the health of crops or domestic or wild animals.

TOXICOLOGY

EPIDEMIOLOGY

The history of beryllium disease is recorded elsewhere in detail (1). Suffice it to say that the acute disease undoubtedly occurred in Germany and Russia during the 1930's but was not widely reported or clearly identified as beryllium-related. The acute disease occurred in Ohio in 1941 among beryllium extraction workers. Chronic disease was first described in the literature in 1946; the patients were Massachusetts fluorescent lamp workers thought to have had tuberculosis or "Salem Sarcoid". (Sarcoid, or more properly sarcoidosis or Boeck's sarcoid, is a granulomatous chronic chest disease of unknown cause. The X-ray may be similar in some respects to tuberculosis, but no infectious organism has been identified. Sarcoidosis and chronic beryllium disease are similar in many ways but are usually distinguishable on the basis of certain clinical and laboratory findings.)

Acute and chronic beryllium disease has been associated with a number of compounds including the fluoride, sulfate, oxyfluoride, oxide, and silicate lattice compound used in phosphors. The metal and beryllium alloys have also been associated with disease. In these latter instances it is not clear whether the metal itself or an oxide is responsible. The metal oxidizes readily on contact with the atmosphere.

There is evidence that the acute disease has been more common than the chronic when the beryllium compound is a soluble acid salt, e.g., sulfate, fluoride. Chronic disease is relatively more common when the compound is an oxide or a phosphor component. Both kinds of disease, however, have been undoubtedly caused by each of these general types of compounds. It has been suggested that all chronic disease is due to oxide present in large or minute amounts and that acute disease is due to acid salts present in large or trace amounts. This may possibly be the case, but the preponderant evidence does not support this interpretation.

There has been controversy for several years as to the relative toxicities of "high-fired" and "low-fired" oxides. There is evidence that the "low-fired" oxides, produced at lower temperatures, are less crystalline and are more active biologically. In animal experiments the "low-fired" materials have relatively higher toxicity, but the "high-fired" materials are by no means inert. It would not appear wise, therefore, to exaggerate the importance of firing temperatures in making judgments about the safety of environmental conditions.

Since low-beryllium alloys (less than 2%) contain relatively low concentrations of toxic metal, it has been stated by some that such alloys are of low hazard. It is prudent, however, to recognize that hazard or lack of hazard is based upon the atmospheric concentrations of the element in any alloy. Poorly controlled machining, or metal working operations which give rise to dust may be hazardous even when "low-beryllium" alloys are used. There are examples of disease to support this point.

Industries in which disease has arisen are listed below:

1. Beryllium extraction
2. Fluorescent lamp production and disposal; phosphor making
3. "Neon" lamp industry
4. X-ray tube industry (beryllium windows)
5. Nuclear energy industry (beryllium moderator, reflector, and neutron source)
6. Alloys: production, metallurgy
7. Metalworking, machining
8. Experimental toxicology
9. Ceramics (BeO refractory)
10. Electron tube production.

So-called "neighborhood cases" of chronic beryllium disease have been observed in persons who have lived in the vicinity of beryllium operations or who have laundered the clothes of beryllium workers. There was no direct industrial exposure in these cases but rather an indirect exposure from plant effluents or from dust in overalls and other work clothes.

Information on the epidemiology of beryllium disease is derived largely from case investigations and follow-ups conducted by the Beryllium Case Registry, an index of all reported cases. The Registry was established in 1951 with the support of the Atomic Energy Commission to examine the etiology and natural course of beryllium-related diseases. Dr. Harriet L. Hardy and her associates (2) have collected and maintained cumulative data on over 700 documented cases. Approximately half the cases are acute and half are chronic. The case fatality rate as recorded in the Registry approximates 5-7% for the acute disease and about 30-35% for the chronic disease. The reported cases represent an unknown fraction of those which may have occurred and gone unrecognized or unreported. At the present time the Registry is maintained at the Massachusetts General Hospital and is supported financially by the U.S. Public Health Service. (See also Chapter 9)

PHARMACOLOGY

Animal studies of toxic beryllium compounds have been conducted in attempts to establish the mechanisms of disease production and the concentrations and characteristics of beryllium compounds known to produce injury. These studies have not been fully satisfactory, but progress has been made in demonstrating the behavior of beryllium in experimental animals. The extent to which these observations can be extrapolated to man remains to be seen.

Among the earliest experimental findings was beryllium-induced rickets in animals. As far as is known, this phenomenon is related to the uptake of calcium and phosphorus from the gastrointestinal tract and has nothing whatsoever to do with man. Under varying conditions of exposure, tumors of the bone and lung have been induced in animals by beryllium administration. While this demonstration of carcinogenicity does not now appear to have a human counterpart, it suggests that the continued surveillance of exposed human populations is important. The potential of various soluble beryllium compounds and oxides prepared under various conditions for causing acute and chronic pulmonary reactions has been recently examined in detail. This work has demonstrated that soluble compounds and low-fired oxides have greater biological significance than do high-fired compounds and beryl.

A number of experimental studies have shown the localization of beryllium in organs, specific cells, and organelles. The binding of beryllium to certain proteins and the influence of beryllium on various enzyme systems has been shown. At the present time, however, it is not possible to present a unified explanation for the pathogenesis of acute or chronic beryllium disease as it occurs in man. The experimental observations while interesting, cannot yet be held to reflect what actually occurs in clinical illness.

While the epidemiological, clinical, and industrial hygiene aspects of beryllium intoxication have been reasonably well defined, the fundamental pharmacodynamics of beryllium compounds remain unknown. There have been a number of investigations which demonstrate the distribution of beryllium after administration by various routes. It is doubtful, however, that the developed data shed much light upon the pathogenesis of beryllium disease. The main point is that compounds of this element are poorly absorbed from the gastrointestinal tract.

Other research has focused upon the influence of beryllium upon the activity of enzymes, primarily those which are magnesium-dependent. Although a number of these enzyme systems are influenced by beryllium in in vitro situations, there is no evidence that these observations are important in the pathogenesis of disease in man or experimental animals. The mechanism whereby beryllium induces neoplastic reactions in animals is not known.

There is evidence that chronic beryllium disease is a manifestation of alterations in defensive immune mechanisms. The clinical course of the disease, its pathology, and its response to therapy tend to support this view. Beryllium does appear to alter the migration of lymphocytes in tissues; however the fundamental pharmacological properties of this element are understood in only the most general and unsatisfactory terms.

CLINICAL ASPECTS

Beryllium disease occurs in acute and chronic forms. Some patients have had both acute and chronic beryllium disease. The pathological mechanisms responsible for acute and chronic disease have not been clearly understood. It does appear, however, that the acute disease is a direct response to irritative substances, while the chronic form is a disorder involving the patient's immune mechanism and may be a type of hypersensitivity reaction. (Hypersensitivity is a term describing a disease mechanism. It is not intended to establish "fault" or some defect on the part of the patient as opposed to a defect in the work environment. This point is relevant in that certain industrial defendants in litigation have attempted to establish that patients are sick because they have some sort of intrinsic dysfunction in their body chemistry.) The view that the mechanisms in acute and chronic disease are dissimilar is well supported but does not represent a unanimous opinion. This point is not particularly relevant to the problem at hand, however.

Acute beryllium disease lasts less than a year, usually several weeks or months, and can affect various parts of the body. Of most importance is acute beryllium pneumonitis, a reaction to certain inhaled compounds. In severe forms it may appear as an insidious or fulminating chemical inflammation of the lungs which may be fatal. Milder forms, and inflammations of the upper airway, nose, and throat are similar in many ways to the common cold. Skin reactions such as dermatitis or superficial skin ulcers are self-limited manifestations of contact with soluble beryllium compounds or the implantation of small crystals or slivers of material in the skin. Current treatment is reasonably effective. These forms of acute disease arise largely out of short-term accidental over-exposures or contaminations.

Chronic beryllium disease, also known as berylliosis (not berylosis, the designation for a non-existent, theoretical pneumoconiosis due to beryl), is a systemic disease due to the inhalation of toxic beryllium compounds. It is marked by a latency or delay in onset, which in some cases has separated the termination of exposure and the onset of symptoms by longer than 20 years. A few patients with long latent periods have relatively mild disease and some may possibly have been "cured" or at least stabilized. Most cases, however, have been marked by progressive pulmonary disease and overwhelming cardiac complications of advanced pulmonary disturbances. Prominent symptoms include shortness of breath (dyspnea) on exercise, cough, and fatigue. The dyspnea may be incapacitating. Some patients suffer from collapse of the lungs (pneumothorax), kidney stones, and complications of prolonged steroid treatment. Tuberculosis and cancer have not been common complications. (Note: beryllium compounds can cause cancer in experimental animals). The prognosis is generally unfavorable in spite of treatment with steroids.

The diagnosis of chronic beryllium disease is established on the basis of the history, clinical pattern, chest X-ray, and laboratory and pulmonary function tests. A skin test is used by some clinicians, but the general view is that it is not of great value. Lung biopsy is a useful diagnostic procedure, but it usually involves open chest surgery and a painful post-operative course. It is accurate to state that most lung biopsies have not been done for the purpose of clinical diagnosis or direct patient benefit. Rather, they have been done to establish exposure and a better case for hearings of workmen's compensation boards.

Assay of beryllium in lung tissue establishes exposure, not the presence of disease. Determinations of beryllium in urine and other body fluids or tissues may establish exposure, but the findings are erratic. When modern techniques have been used, no case of chronic beryllium disease has failed to yield a positive assay for this element in lung tissue.

Incidence of acute and chronic beryllium disease is a function of the quantitative and qualitative characteristics of exposure. The qualitative aspects, as outlined above, are not entirely understood, and the existing evidence is available only in general terms. The quantitative aspects are also not clearly understood although it appears that acute disease may occur when peak transient atmospheric concentrations of beryllium exceed $100 \mu g/m^3$. Under massive exposures the attack rate may be virtually 100%. Air levels of beryllium which may produce chronic disease are not well defined. It does appear, however, that the disease has not occurred when beryllium-in-air concentrations have averaged less than 2 $\mu g/m^3$ over an 8-hour day with no peaks over 25 $\mu g/m^3$. Whether or not chronic disease might eventually occur if the average concentration were 6 or 10 $\mu g/m^3$ is not known.

The attack rate for the acute disease seems to be generally related to beryllium concentration and acute pulmonary dose. The chronic disease has been associated with a less predictable attack rate, approximating 5% or less in the fluorescent lamp industry and a somewhat higher level in certain metallurgical operations. The attack rate may be somewhat dose-related in the chronic disease, but individual person-to-person variables appear to be of greater importance. Some individuals have had both acute and chronic berylliosis.

BERYLLIUM ASSAY AND INDUSTRIAL HYGIENE

Beryllium may be assayed in various samples by colorimetric or physical methods. They are described in detail in texts on the subject (3). The occurrence of the disease is controlled through the application of standard and special air cleaning and ventilation techniques. Periodic monitoring of the air concentration of beryllium is important to the maintenance of acceptable working conditions. Since chronic beryllium disease is, for all practical purposes, incurable, and since its onset may be delayed for years, medical surveillance does not prevent the disease nor represent a valid technique for environmental evaluation.

The following limiting air concentrations for beryllium are those currently recommended:

(a) For workroom air: Threshold limit value of 2 $\mu g/m^3$ as time-weighted average during an 8-hour day. A short-term maximum "ceiling" level of 25 $\mu g/m^3$ is not to be exceeded. "Short-term" is interpreted to be 30 minutes or less.

(b) For community ("neighborhood") air: Threshold limit of 0.01 $\mu g/m^3$, measured at breathing height and averaged over one month. No peaks specified.

REFERENCES

1. Stoeckle, J. D., Hardy, H. L. and Weber, A. L. (1969). Chronic beryllium disease, long-term follow-up of sixty cases. Am. J. Med. 46: 545.

2. Tepper, L. B., Hardy, H. L. and Chamberlin, R. I. (1961). Toxicity of beryllium compounds. Elsevier, Amsterdam.

3. Stokinger, H. E. (ed.). (1966). Beryllium: Its industrial hygiene aspects. Academic Press, New York.

CHAPTER 6. FIVE OF POTENTIAL SIGNIFICANCE

RALPH G. SMITH, Department of Industrial Health,
University of Michigan School of Public Health,
Ann Arbor, Michigan

CHROMIUM

ECOLOGY

Most samples of particulate matter collected from urban air contain a small quantity of chromium as well as a number of other metals. The amount of chromium present will be related to both natural and man-made pollution, for chromium is present in the earth's crust to the extent of about 0.04 per cent. Individual soils and rocks vary greatly in their chromium content, however, and concentrations may range from a few parts per million to greater than 1 per cent. As a result of this wide-spread distribution of the metal, it is also found in water, vegetation, and animal life. Plants frequently contain from 0.1 to 0.5 ppm chromium. The principal ore of chromium is chromite ($FeOCr_2 O_3$) and there are presently no working deposits of the ore in the United States. Chromium has not been singled out for extensive study of its distribution and impact on ecological systems, so that relatively little is known about the transfer of the metal from air to water to living systems, etc. It has been shown, however, to be an essential element in the diet of some animals, particularly laboratory rats, and presumably is also an essential trace element for human metabolism. In this respect, it is similar to other metals which are known to be beneficial in low concentrations and harmful at higher concentrations. There have been no episodes comparable to those for arsenic, selenium, etc., in which farm animals have been poisoned by excessive chromium intake due to environmental contamination, nor is chromium noted to have caused damage to plants of importance in agriculture.

Chromium levels in particulate matter from stations reporting to the National Air Sampling Network have averaged over a number of years slightly more than 0.01 $\mu g/m^3$, while the highest reported value is 0.35 $\mu g/m^3$ (1). Human tissues from individuals without any known exposure to chromium will generally be found to contain a very small concentration of the metal, with levels ranging from 0.1 to 1 ppm on a wet tissue basis (2). Higher levels are, of course, found in the tissues of persons occupationally exposed to chromium compounds. Unlike lead and several other metals, the valence state of chromium is of considerable importance in respect to its toxicity, although it is doubtlessly true that the toxicity of the other metals does vary to a lesser degree with valence. Chromium compounds can exist in several valence states and in additon may exhibit widely varying solubilities. In the case of chromium in urban air, most analyses do not determine the valence state and report only total chromium levels as determined by spectographic or atomic absorption analysis.

Chromium compounds are very useful and are rather widely encountered in industry. Most conspicuous, of course, is the chromium used to plate many objects in order to beautify or protect the base metal from corrosion. Chromium is an important alloying element in steel-making, and stainless steel production accounts for a large consumption of chromium. Many pigments contain chromate compounds, chromite ore is useful as a refractory material, chrome tanning of leather is widely practiced, and chromate compounds are used as water additives to prevent corrosion. Wide-spread pollution of the atmosphere by chromium compounds is therefore possible, and a certain percentage of the urban chromium pollutant levels undoubtedly originates from industrial sources. The local air pollution problem created by the emission of chromic acid mist from plating plants is well-known, and results in frequent complaints of damage to metallic surfaces. The use of chromates for the prevention of corrosion in cooling water is probably a significant source of chromate emission to the atmosphere and with the trend toward increasing use of water-cooling towers for power generating installations, may constitute a problem of growing magnitudes.

TOXICOLOGY

It is probably true that all compounds of chromium are toxic in sufficiently high concentrations, but certain compounds have been demonstrated to be much more toxic than others. In general, the metal itself is very nearly inert and does not constitute an air pollution problem. Divalent compounds have not generally been considered to be toxic, but recent evidence indicates that workers in the chromate industry may be exhibiting evidence of lung damage due to trivalent chromium. By far the most important effects of exposure to chromium result from the hexavalent compounds, or chromates. Depending upon their solubility to some extent, these compounds are usually irritating and toxic to all tissues. Of the greatest concern with respect to industrial exposure and air pollution exposure is the reported high incidence of cancer of the lung resulting from exposure to hexavalent chromium compounds (3,4,5). Although there is some doubt about the exact compounds which produce cancer in industrial workers, it is generally accepted that hexavalent compounds are at least partially responsible, and some animal studies have tended to support the evidence based on the epidemiological studies of workers exposed to chromium compounds. Cancer in the respiratory tract is the only known type of cancer related to chromium compounds and in the great majority of cases this is found to be bronchogenic carcinoma. Cancers have also been noted in the upper respiratory tract, however, but not elsewhere.

The immediate effects of exposure to chromium compounds depend greatly upon the properties of the substance involved, and doubtlessly the most frequently observed effect is that due to the inhalation of chromic acid mist in chromium plating. This mist, which contains chromate compounds as well as sulfuric acid, is intensely irritating and will cause damage to the upper respiratory system if over-exposure takes place. Classically, chrome workers exhibit perforation of the nasal septum, ulceration of the septum, and a variety of other effects attributable to irritation of the respiratory system. The incidence of the symptoms and a description of them may be found in the report of a large scale study (6) conducted by the United States Public Health Service in which 897 chromate workers were examined. Except for the effects on the respiratory system, chromates are not thought to cause systemic toxic effects.

The most wide-spread complaint resulting from exposure to chromates is the production of dermatitis which, in some cases, appears to result from sensitization of individuals who thereafter may exhibit inflammation of the skin on the hands, arms, or other parts of the body. Portland cement may contain several parts per million chromate compounds and contact with cement has resulted in dermatitis appearing in sensitized individuals (7). When chromate compounds contact the skin directly, ulcers may be formed, known in some industrial situations as "chrome holes" (8). Such ulcers may appear on the hands or in areas where dust can accumulate and may be very painful and slow to heal.

The threshold limit values (TLV's) recommended by the American Conference of Governmental Industrial Hygienists for chromium compounds are dependent upon the form in which the chromium exists (9). For chromic acid and chromates expressed as CrO_3, the TLV is 0.1 mg/m^3. For other soluble chromic and chromous salts, the TLV is 0.5 mg/m^3 expressed as chromium. For the metal and its insoluble salts the TLV is 1 mg/m^3. These values represent concentrations which are presumably tolerable for a normal work week throughout the lifetime of those occupationally exposed. The effects on human health previously summarized are presumably related to concentrations as high or higher than the TLV's and are not going to result from environmental exposure to a few hundredths of a $\mu g/m^3$. As in the case of most trace metals, little or nothing is known of the effects of exposure to environmental chromium compounds at these levels and, thus far, no air quality standard has been set for chromium compounds. The USSR has adopted a value of $1.5 \ \mu g/m^3$ as the maximum permissible average daily concentration for chromates, and 80 μg/m³ for average 24-hour exposure to trivalent chromium and its compounds. If an air quality standard is adopted for chromium, it can be expected to be very much lower than the TLV, but conclusive evidence for a standard based on health effects is presently lacking. It would appear to be logical to set a rather stringent standard for hexavalent chromium in view of industrial experience with its ability to produce lung cancer.

The major sources of chromate pollution, such as plating plants and industries which emit particulate chromium compounds to the air, can readily be controlled by available technology, and usually some degree of control presently exists due to the irritating and corrosive nature of such compounds. The contribution to the atmosphere from the burning of coal, however, is controlled by measures presently designed to remove particulate matter and will continue to be reduced as the permissible standards of emission of particulate matter are lowered and enforced. It is probable that chromates in water cooling facilities will pose a problem requiring further consideration in some areas, while the general use of chromate-containing pigments, etc., will continue to provide a rather wide-spread and diffuse source of chromates which ultimately find their way into the environment. If it should develop that chromates are of greater concern than is presently known to be the case, control measures will undoubtedly include efforts to minimize the use of such compounds and the substitution of other materials thought to be less objectionable.

REFERENCES

1. Sullivan, R. J. (1969). Air Pollution Aspects of Chromium and Its Compounds. Technical Report for National Air Pollution Control Administration, DHEW, Litton Industries, Environmental Systems Division, Bethesda, Maryland.

2. Baetjer, A. M. (1956). Relation of chromium to health. In: M. J. Udy (ed.), Chromium. Vol. 1, Am. Chem. Soc. Monograph 132. Reinhold, New York.

3. Baetjer, A. M. (1950). Pulmonary carcinoma in chromate workers. Arch. Ind. Hyg. Occup. Med. 2: 487.

4. Bidstrup, P. L. and Case, R. A. M. (1956). Carcinoma of the lung in workmen in the bichromates-producing industry in Great Britain. Brit. J. Ind. Med. 13: 260.

5. Mancuso, T. F. and Hueper, W. C. (1951). Occupational cancer and other health hazards in a chromate plant; a medical appraisal. 1. Lung cancer in chromate workers. Ind. Med. Surg. 20: 359.

6. Gafafer, W. M. et al. (1953). Health of workers in chromate producing industry. U. S. Public Health Service Publication No. 192.

7. Cairns, R. J. and Colman, C. D. (1962). Green tattoo reactions associated with cement dermatitis. Brit. J. Dermatol. 74: 288.

8. Stokinger, H. E. (1963). Chromium. In: F. A. Patty (ed.), Industrial Hygiene and Toxicology, Vol. 2, (2nd ed.). Interscience, New York, 1017.

9. Threshold Limit Values of Airborne Contaminants. (1970). American Conference of Governmental Industrial Hygienists, Cincinnati, Ohio.

MANGANESE

ECOLOGY

Manganese is a relatively common metal, occurring in the earth's crust at sufficiently high concentrations to make it the 12th most abundant element. It occurs in a number of different ores, but relatively little manganese is mined within the United States, and hence air pollution problems related to mining are not a domestic problem. Most of the manganese imported (more than 90%) is used in the iron and steel industry, so that the potential for pollution of the atmosphere is large. Manganese is almost always found in the particulate matter of urban air, and National Air Sampling Network data show that concentrations may range from a low of several hundredths $\mu g/m^3$ to more than 10 $\mu g/m^3$ (1). Maximum daily averages frequently fall in the range of several $\mu g/m^3$. In the manufacture of steel, if total particulate emissions are not controlled, some manganese will inevitably be present, and in a report of the air pollution aspects of the iron and steel industry (2) it was stated that as much as 4% of the fume emitted by an electric arc furnace consisted of manganese oxide. The production of ferromanganese, which is required in making steel, is a source of much higher concentrations of atmospheric manganese and historically has resulted in damage to the health of those in communities surrounding several such plants. The relatively high concentration of manganese in the earth's crust inevitably results in this element being found in fossil fuels. In studies of emissions from coal-fired power plants (3), manganese concentrations ranging from 60-400 $\mu g/m^3$ have been reported. Fuel oil also contains manganese and the burning of residual fuel oil may give rise to concentrations of manganese as high as 55 $\mu g/m^3$ (4). Manganese compounds find many additional uses, but the use of organic manganese compounds as fuel additives is obviously of the greatest interest in respect to atmospheric levels of manganese. One such compound, methylcyclopentadienyl manganese tricarbonyl, is used along with tetraethyl lead as a gasoline additive.

Manganese is an essential element for some plant life and, hence, is included in certain fertilizers, generally as manganese ethylene-bis-dithio-carbamate. Some air pollution may result from this usage and, in addition, of course, substantial quantities of manganese are introduced into the soil and the environment in general. Manganese compounds have long been known to be effective catalysts for certain reactions and the commercial preparation known as hopcalite is a mixture of manganese and other metallic oxides which can be used to catalyze the oxidation of carbon monoxide or hydrocarbons at elevated temperatures. This same catalytic ability makes manganese of particular interest in respect to its air pollution capability, for there is evidence to show that manganese compounds such as MnO_2 in the air act to catalyze the oxidation of sulfur dioxide to sulfur trioxide, and ultimately to sulfuric acid. Thus, the importance of manganese concentrations in the air becomes a function not only of its direct effects but also of the general problem of sulfur oxides in the air and the effects on health and materials realted to them.

TOXICOLOGY

Manganese has long been known to be a toxic element, and illness resulting from occupational exposure to manganese and its compounds has been well-documented in the scientific literature. Considerations of toxicity are complicated by the fact that manganese can exist as compounds in up to eight different valence states. These, in turn, may be cations such as Mn^{++} or anions such as MnO_4^-. Stokinger (5) has summarized much of the available information concerning the toxicity of the numerous compounds of manganese and, in general, it appears that cations are more toxic than anions, and that Mn^{++} is more toxic than Mn^{+++}. Inasmuch as manganese determinations in the atmosphere are usually total manganese without knowledge of the form in which it is present, it is difficult to apply such information to the problem of the health effects of air-borne manganese.

The principal effects of long-term occupational exposure to manganese compounds are the production of manganese poisoning and manganese pneumonia. Manganese is similar to mercury in exerting its primary effect on the central nervous system, with the resulting disease proportional in severity to the length and intensity of the exposure. Early symptoms include languor, sleepiness, and complaint of weakness in the legs. A mask-like facial expression may develop, and muscular twitching and tremor of the hands progressing in more severe cases of the arms and legs may occur. Emotional disturbances may be noted, including uncontrollable laughter, speech difficulty, hallucinations, mental confusion, insomnia, and similar types of abnormal behavior. If a diagnosis of manganese poisoning is made soon enough, the disease is usually reversible, providing the individual is removed from further exposure to the dust. In addition to, or in place of manganese poisoning as just described, a pneumonia may result from manganese exposure which is characterized by

146

high temperature and dyspnea among other symptoms, and which does not respond to antibiotics. This disease has actually been reported as a consequence of gross overexposure to air pollution, and occurred more than 35 years ago in a small town in Norway (6). An industrial plant in this town, engaged in the manufacture of manganese alloys, emitted large quantities of manganese dust into the air, and subsequent studies of the townspeople showed that mortality due to pneumonia in that area was greatly in excess of that for the rest of Norway. In addition, analyses of lung tissue from those who died revealed that the tissue contained considerably more manganese than normal lungs. Sullivan (7) has summarized similar findings from surveys made in the USSR, and in Italy. It should be noted that Heine (8) alleged that all cases of so-called manganic pneumonia such as those just cited can be explained on the basis of such other factors as undernourishment, or drought, etc., and he was further unable to produce the disease in animals experimentally. Stocks (9) also was unable to find any correlation between pneumonia and manganese in urban air although such a correlation was noted between pneumonia and beryllium. There appears to be no evidence at the present time to indicate that the low concentrations of manganese which are generally found in urban air have produced any form of manganese poisoning, or increased incidence of pneumonia.

The control of atmospheric manganese levels, like that of other trace metals is basically being achieved by measures designed to reduce levels of particulate matter in the air. In general, there are no problems unique to manganese-bearing particles, and conventional dust and fume collecting devices are considered adequate for the task. As previously noted, the use of organic manganese additives to gasoline is of particular interest and will, no doubt, be singled out for special consideration by air pollution control authorities.

REFERENCES

1. Air Quality Data from the National Air Sampling
    Network and contributing State and local
    networks, 1964-1965. (1966). U. S. Dept. of
    Health, Education and Welfare, Public Health
    Service, Cincinnati, Ohio.

2. Schueneman, J., High, M. D. and Bye, W. E.
    (1963). Air pollution aspects of iron and steel
    industry. U. S. Dept. of Health, Education
    and Welfare, Public Health Service Publication
    No. 999-AP-1.

3. Smith, W. S. and Gruber, C. W. (1966).
    Atmospheric emissions from coal combustion -- an
    inventory guide. U. S. Dept. of Health,
    Education and Welfare, Public Health Service
    Publication No. 999-AP-24.

4. Smith, W. S. (1962). Atmospheric emissions
    from fuel oil combustion -- an inventory guide.
    U. S. Dept. of Health, Education and Welfare,
    Public Health Service Publication No. 999-AP-2.

5. Stokinger, H. E. (1963). Manganese, In:
    F. A. Patty (ed.), Industrial Hygiene and
    Toxicology, Vol. 2, (2nd ed.). Interscience
    Publishers, New York, pp. 1079.

6. Elstad, D. (1939). Manganholdig Fabrikkrevek
    som Medvirkende Arsak ved Pneumoni-Epidermier i
    en Industribygd. Nord. Med. 3: 2527.

7. Sullivan, R. J. (1969). Air Pollution Aspects
    of Manganese and Its Compounds. Technical Report
    for National Air Pollution Control
    Administration, DHEW, Litton Industries,
    Environmental Systems Division, Bethesda,
    Maryland.

8. Heine, W. (1944). Observations and experimental
    investigation of manganese poisonings and
    manganese pneumonia. Z. Hyg. Infektions
    Krankheiten 125: 1.

9. Stocks, P. (1960). On the relationship between
    atmospheric pollution in urban and rural
    localities and mortality from cancer, bronchitis

and pneumonia, with particular reference to 3, 4 benzopyrene, beryllium, molybdenum, vanadium, and arsenic. Brit. J. Cancer 14: 397.

NICKEL

ECOLOGY

In a study of the nickel content of suspended particulate matter in 30 cities of the United States, McMullan (1) reported that the mean nickel level was 0.045 $\mu g/m^3$ for the four-year period ending in 1960, and 0.35 $\mu g/m^3$ for the next four years. On the average, these nickel levels constituted about 0.3 % of the total suspended particulate matter. As in the case with most other trace metals, these levels represent a combination of background nickel from natural sources and nickel discharged by the activities of man. It is probable, however, that in the case of nickel, a larger percentage of the metal in the airborne particulate matter is industrial rather than natural in origin. Schroeder et al. (2), have reported finding nickel in virtually every kind of sample examined, including water, plants, and animal tissues. The burning of coal is a source of substantial quantities of nickel in the atmosphere and coal ash has been shown to contain nickel in amounts varying from 3 ppm to 10,000 ppm, depending upon the origin of the coal (3). Crude oil may also contain a substantial amount of nickel and it has been reported by Given (4) that a sample of crude oil contained 110 ppm nickel. The burning of heavy fuel oils results in fly ash which can be quite rich in nickel, and in two reported tests the nickel oxide content was 1.8 and 13.2 % (5). Diesel oil contains far less nickel, Frey and Corn (6) having reported 2 ppm in the oil itself, while particulate matter samples collected in the exhaust resulting from the burning of diesel oil contained as much as 1 % nickel. There have been no reports of wide-spread damage to the environment by nickel compounds in the atmosphere or elsewhere, and nickel does not appear to be a problem with respect to its effect on vegetation, domestic animals, or wild life. There has been concern over the effect of atmospheric nickel levels on the health of humans, however, and this concern is based on well-documented evidence that certain nickel compounds are known to be carcinogenic to man and animals. Nickel and its compounds are also toxic in the conventional sense, but the levels required to produce observable effects are considerably greater than those found in ambient air.

TOXICOLOGY

Most of our knowledge concerning the toxicity of nickel compounds is based upon evidence of two kinds, that resulting from industrial exposure to nickel and that resulting from animal experimentation. Although poisoning by nickel and its salts is very rare, the salts are toxic if introduced directly into the blood stream. The threshold limit value (TLV) established by the American Conference of Governmental Industrial Hygienists (7) is 1 mg/m^3. On a comparative basis, this would indicate that nickel is less toxic than lead (TLV 0.2 mg/m^3) but, as indicated previously, chronic poisoning by nickel compounds is a relatively rare event. One of the most common exposures to nickel compounds is that resulting from electroplating with nickel and the most wide-spread problem is the production of a dermatitis popularly known as the "nickel itch." The dermatitis usually appears at the area of contact, while a more wide-spread reaction is frequently observed which may be indicative of an allergic response to the metal. Some persons even suffer from contact with nickel plated objects and several studies have been published in which individuals were non-occupationally exposed to nickel plated objects with the subsequent development of nickel dermatitis. The most toxic of all nickel compounds, and probably one of the most toxic compounds used in industry, is nickel carbonyl, a low boiling liquid which readily vaporizes at ambient temperatures and which decomposes at temperatures in excess of 60 C. The TLV for nickel carbonyl is 0.001 ppm, making it the most toxic gas on the entire list of substances for which TLV's have been set. The most important consequences of exposure to nickel carbonyl, however, are not those resulting from its toxicity in the usual sense, but rather the production of cancer of the lung. Sunderman, Sunderman, Jr., and their co-workers have studied the carcinogenic action of nickel carbonyl and report their findings in a series of papers (8,9,10) which show conclusively that both animals and humans develop pulmonary cancer as a result of exposure to nickel carbonyl. A number of other studies, which have been summarized by Sullivan (11), clearly tend to implicate nickel carbonyl as a carcinogen in industry. Certain other nickel compounds are also thought to be carcinogenic by some, and many studies have shown that nickel workers exhibit substantially higher-than-expected incidence of pulmonary and nasal cancer. In view of the fact that exposure to nickel in industry is usually complicated by concomitant exposure to other metals, some controversy exists concerning the true carcinogenicity of nickel compounds and the chemical form in which these compounds are most active.

Although cases of cancer or illness attributable to environmental nickel compounds have not been reported, it is logical to investigate the possibility that the reported higher incidence of lung cancer in urban populations might in some way be related to nickel levels. Sunderman and Sunderman, Jr. (12), in tests conducted upon cigarette smoke, have expressed the opinion that there is sufficient nickel in cigarettes to produce nickel carbonyl in quantities which could be responsible for the production of lung cancer in smokers. Whether or not ambient levels of nickel are sufficient to give rise to increases in the incidence of lung cancer, however, is a matter of conjecture at this time and no experimental evidence exists which might resolve the issue.

No environmental air quality standard has been set for nickel or its compounds to date and the only existing standards are those previously cited as TLV's for the control of industrial exposure. Except in the special case of proximity to an industrial plant using nickel carbonyl, it is improbable that this form of nickel would be encountered as an air pollutant. Published data relate to the nickel content of particulate matter, however, and it is unlikely that extensive sampling for nickel carbonyl has been performed. It is known that nickel carbonyl can be formed quite readily when sufficiently high concentrations have occurred where this toxic gas has been unexpectedly encountered. Potentially, one of the more important sources of nickel in the atmosphere could be the burning of fuels containing nickel organic compound additives. A number of compounds such as cyclopentadienyl nickel complexes and nickel organic phosphates have been proposed as fuel additives, but in view of the effects of nickel previously referred to, it is obvious that the wide-spread use of such substances should be undertaken only if it can be demonstrated that such use does not produce atmospheric levels of nickel harmful to man or the environment.

REFERENCES

1. McMullen, T. B. (1969). Concentrations of nickel in urban atmosphere (1957-1964). In: Air Pollution Aspects of Arsenic and Its Compounds. Technical Report for National Air Pollution Control Administration, DHEW, Litton Industries, Environmental Systems Division, Bethesda, Maryland.

2. Schroeder, H. A., Balassa, J. J. and Tipton, I. H. (1962). Abnormal trace metals in man -- Nickel. J. Chronic Dis. 15: 51.

3. Abernethy, R. F. and Gibson, F. H. (1963). Rare elements in coal. Bur. Mines Inform. Circ. IC-8163.

4. Given, P. H. (1966). Coal science. Adv. Chem. Ser. 55.

5. Smith, W. S. (1962). Atmospheric emission from fuel oil combustion -- an inventory guide. U. S. Dept. of Health, Education and Welfare, Public Health Service Publication No. 999-AP-2.

6. Frey, J. W. and Corn, M. (1967). Physical and chemical characteristics of particulates in a diesel exhaust. Am. Ind. Ass. J. 28: 468.

7. Threshold Limit Values of Airborne Contaminants. (1970). American Conference of Governmental Industrial Hygienists, Cincinnati, Ohio.

8. Sunderman, F. W. and Donnelly, A. J. (1965). Studies of nickel carcinogenesis: Metastasizing

pulmonary tumors in rats induced by the inhalation of nickel carbonyl. Am. J. Pathol. 46: 1027.

9. Sunderman, F. W. et al. (1957). Nickel poisoning. 4. Chronic exposures of rats to nickel carbonyl; A report after one year observation. Arch. Ind. Hyg. 16: 480.

10. Sunderman, F. W. et al. (1959). Nickel poisoning. 9. Carcinogenesis in rats exposed to nickel carbonyl. Arch. Ind. Hyg. 20: 36.

11. Sullivan, R. J. (1969). Air Pollution Aspects of Nickel and Its Compounds. Technical Report for National Air Pollution Control Administration, DHEW, Litton Industries, Environmental Systems Division, Bethesda, Maryland.

12. Sunderman, F. W. and Sunderman, Jr., F. W. (1961). Nickel poisoning. 11. Implication of nickel as a pulmonary carcinogen in tobacco smoke. Am. J. Clin. Pathol. 35: 203.

VANADIUM

ECOLOGY

Vanadium compounds have long been known to be present in fossil fuels such as coal and oil at concentrations substantially greater than that of many other so-called trace metals. In a report compiled by Athanassiadis (1) the vanadium content of coal from various sections of the United States is shown to vary from 16 to 176 ppm, while the resulting ash is enriched to such an extent that levels as high as 1000 ppm (0.1%) are found, depending on the region from which the crude was taken. When residual oils are burned, the ash may be very rich in vanadium, with concentrations reported to range from a few per cent, to as high as 63% (expressed as V_2O_5).

In view of the enormous quantities of coal and oil which are burned daily, it is not surprising that the particulate matter in urban air always contains a measurable amount of vanadium, even though the element is one of the less abundant elements in the earth's crust. The air of most cities contains a fraction of a microgram of vanadium per cubic meter, and maximum concentrations in excess of 1 $\mu g/m^3$ have been reported (2). It is probable that substantially higher levels would be found if sampling were conducted at appropriate locations downwind of installations burning vanadium-rich fuels, but little data have been published. As a result of the airborne dissemination of vanadium, there must be a continuing deposition of this element on the waters and land masses of the earth, but no reports of damage to the ecosystem have appeared in the literature.

TOXICOLOGY

Concern has been expressed that health may be adversely affected due to vanadium levels, however, and several epidemiological studies have shown correlations between air levels of vanadium and some health parameters. It is difficult to draw unequivocal conclusions from these studies because the actual exposures were to all of the air pollutants present in urban air and not simply to vanadium. However, Hickey (3) in a statistical analysis of data on the concentration of 10 metals in 25 communities in the United States did report that vanadium levels showed a strong correlation with diseases of the heart. The actual findings were quite complex, and showed interrelationship between vanadium and certain other metals, notably nickel and cadmium. Stocks (4) using a similar approach with thirteen trace elements reported correlations with several respiratory diseases, including cancer of the lung. Actually, vanadium levels appeared to correlate strongly with bronchitis in males, and also with pneumonia in males and females, but the association with lung cancer was weak.

To some extent the statistical study findings are consistent with what is known of the behavior of vanadium at higher concentrations, although there is no direct evidence that vanadium is a carcinogen to man or experimental animals. Industrial exposure to high concentrations of vanadium, usually as the pentoxide, has long been known to result in pulmonary irritation, as well as irritation of the eyes and upper respiratory system. In a summary of industrial experience with vanadium compounds, Stokinger (5) notes that workers may suffer from cough, irritation of the conjunctiva, nasal catarrh, dryness and irritation of the throat, wheezing and dyspnea. In one study cited, several Swedish workers who were examined eight years after initial exposure were found to be suffering from bronchitis and other symptoms, apparently as the result of a chronic disease caused by previous overexposure. A number of reports are also cited by Stokinger, in which the production of a mild, chronic inflammatory change in the respiratory tract is attributed to cleaning oil-fired burners or gas turbines.

Extensive biochemical studies have been published concerning the action of vanadium, and much is known of the effect of vanadium compounds on the synthesis of cholesterol, the inhibition of acetylcholinesterase and a variety of other biochemical processes. The effect on cholesterol synthesis has been most widely studied and it is well established that vanadium does inhibit the synthesis of cholesterol. It has been demonstrated more specifically that vanadium inhibits the activities of certain coenzymes involved in the early stages of cholesterol synthesis as well as the formation of mevalonic acid, which is an intermediate in the scheme of cholesterol synthesis. Vanadium also seems to exert an effect on existing stores of cholesterol and, in addition, tends to reduce the retention of cholesterol in the diet.

In common with organic phosphate compounds and certain other substances, vanadium also can inhibit the activity of the enzyme cholinesterase. One of the consequences of such inhibition is the tendency to produce a deficiency in choline concentrations. The consequences of choline deficiency, as summarized by Athanassiadis, include choline deficiency, which is known to be related to fatty infiltration of the myocardium of the heart, a decrease in serum albumin, and adverse effects in general relative to reproduction and pregnancy. Other effects attributable to choline deficiency involve fatty degeneration and cirrhosis of the liver, damage to the kidney, anemia and muscular dystrophy.

Athanassiadis has speculated that, inasmuch as vanadium is known to be an inhibitor of cholinesterase, it may indeed be correlated with some of these choline deficiency effects. Other biochemical substances which may be affected by excessive vanadium levels include the adrenocortical hormones, and the sulfur containing amino acids cystine, cysteine, and methionine.

An extensive literature exists involving animal exposure studies to vanadium compounds and, in general, these studies have formed a basis for some of the observations just summarized, and in addition, have served to establish lethal doses and data concerning the specific toxicity of vanadium compounds. Toxicity appears to be dependent, to a certain extent, upon the valence of the vanadium compounds, and pentavalent compounds such as V_2O_5 are generally more toxic than trivalent compounds. In recognition of the toxicity of vanadium compounds, the American Conference of Governmental Industrial Hygienists has established threshold limit values (TLV's) for vanadium as follows (6). For vanadium pentoxide (V_2O_5) as a dust the TLV is 500 $\mu g/m^3$. For vanadium pentoxide as a fume the TLV is 100 $\mu g/m^3$ The lower TLV for fume recognizes that the particles tend to be much finer and more readily inhaled and absorbed than do dust particles. The comparable industrial limits established by the USSR are identical to those established by the ACGIH except that the diminished toxicity of ferrovanadium and vanadium carbide are recognized by permitting substantially higher air concentrations of these substances. Thus far no environmental air standards have been set for vanadium or its compounds in the general atmosphere.

There are a number of industrial sources of vanadium in addition to the countless installations which burn coal or residual fuel oil. In one report (7), a total of 119 industrial units which produce major vanadium chemicals is listed with a large number of these being located in New Jersey and New York. Most of the vanadium produced annually is used in the manufacture of steel, but lesser amounts are used in a variety of ways, such as the catalysis of many important chemical reactions in industry. Vanadium pentoxide is also being used to an increasing extent as a catalyst in the manufacture of sulfuric acid by the contact process, replacing the more expensive platinum previously used for this purpose. It seems likely that the attainment of this standard will require a considerable amount of reduction in the emission of vanadium compounds to the atmosphere. In view of the many sources of this metal, such a program will have wide-spread impact, but as in the case of many other metals and fuels, measures taken to reduce total particulate emissions will inevitably reduce vanadium emissions also.

REFERENCES

1. Air Pollution Aspects of Vanadium and Its Compounds. Technical Report for National Air Pollution Control Administration, DHEW, Litton Industries, Environmental Systems Division, Bethesda, Maryland (Sept. 1969).

2. Air Quality Data from the National Air Sampling Network and contributing State and local Networks, 1964-1965. (1966). U. S. Dept. of Health, Education, and Welfare, Public Health Service, Cincinnati, Ohio.

3. Hickey, J. R., Schoff, E. P. and Clelland, R. C. (1967). Relationship between air pollution and certain chronic disease death rates -- multivariate statistical studies. Arch. Env. Health 15: 728.

4. Stocks, P. (1960). On the relation between atmospheric pollution in urban and rural localities and mortality from cancer, bronchitis, and pneumonia, with particular reference to 3, 4 benzopyrene, beryllium, molybdenum, vanadium, and arsenic. Brit. J. Cancer 14: 397.

5. Stokinger, H. E. (1955). Organic beryllium and vanadium dusts. Arch. Ind. Health 12: 1171.

6. Threshold Limit Values of Airborne Contaminants. (1970). American Conference of Governmental Industrial Hygienists, Cincinnati, Ohio.

7. Directory of Chemical Producers. (1968). Stanford Research Institute, Menlo Park, California.

ARSENIC

ECOLOGY

Virtually any sample of the environment will be found to contain a small quantity of arsenic, provided that a sample of sufficient size is analyzed by a suitably sensitive method. In most cases, the levels of arsenic found are essentially "background" levels, and are not properly considered as part of man's pollution input. There are, of course, many notable exceptions in the literature which give ample proof that arsenic pollution was and, to some extent, is a reality; but in such cases, levels much higher than background are usually encountered. Most soils will contain several ppm arsenic, while sea water will average several ppb. Fresh waters vary greatly, depending on the water sources and geographical location, but even rainwater has been reported to contain as much as 14 ppb arsenic. Air is almost always found to contain a small quantity of arsenic in the particulate matter, and except for some western cities and certain other areas where there are local sources of arsenic-containing discharges, levels generally are quite low, usually several hundredths $\mu g/m^3$ of air.

Although sea water contains relatively little arsenic, seafood is usually rich in this element, demonstrating once again the remarkable enrichment of a trace substance by living creatures in equilibrium with their environment. Schrenk and Schreibeis (1) have reported, for example, that oysters contain 3-10 ppm, lobsters 70 ppm, mussels as much as 120 ppm, and prawns up to 170 ppm. It is apparent that a seafood diet can easily be a major source of arsenic intake, and these authors demonstrated conclusively that a good seafood dinner resulted in prompt elevation of urinary arsenic levels, a factor of some importance when attempting to assess occupational exposure by urinary arsenic determinations.

It does not appear that any impairment of human health, or damage to the ecosystem in general, results from the arsenic levels just cited. In contrast to these background levels, or perhaps even urban levels where no concentrated discharges are taking place, several community episodes have been reported, in which large amounts of arsenic were distributed over areas adjacent to plants engaged in smelting ores containing arsenic. In one of the earliest reported episodes (2), animals grazing on plants contaminated with arsenic trioxide were killed, while more recently injury to humans was reported in Chile (3), and in one of the western states of the United States (4).

In the past, and to a lesser extent today, arsenical compounds have been used as pesticides, giving rise to very high soil levels, and, during the time of application and at certain other times, high air levels. In a thorough study (5) of arsenic exposure in the Wenatchee, Washington apple orchards in 1938, air levels of several hundred, and even several thousand $\mu g/m^3$, were reported while application of the pesticides was taking place. In addition, of course, arsenic residues on the fruit were of concern, and provided another source of intake for the population-at-large.

Another agricultural use of organic arsenicals is the defoliation of cotton plants prior to picking the cotton by machine, and elevated atmospheric levels have been reported in the vicinity of cotton gins.

TOXICOLOGY

All compounds of arsenic are toxic, and in sufficient concentration will cause death or illness in humans, animals, or plants. The particulars of acute poisoning by arsenic compounds due to ingestion or inhalation are of little interest in relation to environmental pollution, but in summary many vital functions are affected, and death may result from relatively small amounts of arsenic. Chronic intoxication due to ingestion or inhalation has been well described by Dinman (6) and, unlike the acute disease, early manifestations include subjective complaints and such non-specific symptoms as weakness, malaise, abdominal complaints, and pains involving the extremities. The most common ailments in the case of those occupationally exposed to arsenic are dermatitis, perforation of the nasal septum, and even ulceration of the skin. Many other symptoms have been noted as a result of chronic

159

exposure to arsenic but the probability of their occurrence as a result of environmental exposure is very slight.

Arsenic is usually included on the list of substances suspected of being carcinogenic to humans. Hueper (7), for example, considers arsenic to be a recognized carcinogen, with the skin, lung, and liver as recognized sites of cancer. There is also evidence that lung cancer is more prevalent in certain occupational groups where arsenic exposure is involved, but counter arguments are based on the failure to note cancer in other occupational groups known to be exposed to arsenic. A number of references are cited by Sullivan (8) which argue for and against the likelihood that arsenic in the industrial environment is carcinogenic, and although the matter is not resolved, clearly this potential carcinogenicity is perhaps the single most important aspect of concern with arsenic in the environment.

Arsenic compounds have been used for many years in the treatment of certain diseases and, in particular, syphilis, but antibiotics have largely replaced the arsenical drugs, so that they are not in wide use today. One of the most toxic compounds of arsenic is the gas, arsine, which may be encountered accidentally whenever nascent hydrogen comes into contact with a solution containing arsenic. Arsine is extremely toxic and a single short exposure to a concentration of only 100 ppm may be sufficient to cause death or serious illness. Its principal effect is the destruction of red blood cells resulting in a hemolytic anemia or jaundice. It is highly unlikely that arsine would become an air pollutant of concern to those not occupationally exposed to it, and the literature contains no evidence of such exposure.

Some measure of the toxicity of arsenic compounds may be gained by noting the threshold limit value (TLV) established by the American Conference of Governmental Industrial Hygienists for occupational exposure (9). The TLV for arsenic and its compounds expressed as arsenic is 0.5 mg/m^3. Presumably those exposed occupationally to arsenic compounds may tolerate this concentration, calculated as a time-weighted average exposure, throughout their occupational lifetime without the development of symptoms of arsenic intoxication. The comparable TLV for arsine is 0.2 mg/m^3 or 0.2 ppm as it is ordinarily expressed. This value makes arsine one of the most toxic gases on the entire list of substances for which TLV's have been set. (The TLV for carbon monoxide, for example, is 50 ppm.) At the present time, no air quality standards for arsenic have been proposed, but very likely, such standards will be very much lower than the TLV's just cited. Both the USSR and Czechoslovakia have established 24-hour standards of 3 $\mu g/m^3$ for arsenic and its compounds (9).

It does not appear that arsenic constitutes a growing threat to the environment and, indeed, may be an excellent example of a toxic substance whose emission to the atmosphere and to the environment in general has undergone a substantial reduction during recent years. In addition to abatement procedures which have been applied to industrial sources of arsenic, the factor most responsible for diminishing the amount of arsenic in the environment, has been the replacement of arsenical pesticides with others which have been found to be more effective. While it may be true that the substitute pesticides have in turn created serious problems in the environment, they have at least been responsible for reducing the input of a persistent substance which will never undergo any biodegradation. Further control measures will doubtlessly be required when air quality standards are adopted and will consist of reducing the emission of airborne arsenic compounds from industries already discussed. Although the burning of coal constitutes a major source of arsenic, on a tonnage basis, it does not appear likely at the present time, that modifications in the methods of burning coal will be required solely to minimize the emission of arsenic. Arsenic levels will no doubt tend to be reduced when other steps are taken to minimize the overall pollution caused by coal burning. It is always possible that continuing research will bring to light hitherto unsuspected problems, but the magnitude of arsenic pollution is known and, in general, adequately controlled. Should further evidence relating ambient levels of arsenic to human lung cancer be produced, there could arise a very great interest in reducing arsenic levels, but to date no such evidence has been forthcoming.

REFERENCES

1. Schrenk, H. H. and Schreibeis, L. (1958). Urinary arsenic levels as an index of industrial exposure. Amer. Ind. Hyg. Assoc. J. 19: 225.

2. Haywood, J. K. (1907). Injury to vegetation and animal life by smelter fumes. J. Am. Chem. Soc. 19: 998.

3. Oyanguren, H. and Perez, E. (1966). Poisoning of industrial origin in a community. Arch. Env. Health 13: 185.

4. Birmingham, D. J. et al. (1965). An outbreak of arsenical dermatosis in a mining community. Arch. Dermatol. 91: 457.

5. Neal, P. A. et al. (1941). A study of the effect of lead arsenate exposure on orchardists and consumers of sprayed fruit. U. S. Public Health Service, Public Health Bulletin No. 267.

6. Dinman, B. D. (1960). Arsenic: Chronic human intoxication. J. Occup. Med. 2: 137.

7. Hueper, W. C. (1963). Environmental carcinogenesis in man and animals. Ann. N. Y. Acad. Sci. 108: 963.

8. Sullivan, R. J. (1969). Air Pollution Aspects of Arsenic and Its Compounds. Technical Report for National Air Pollution Control Administration, DHEW, Litton Industries, Environmental Systems Division, Bethesda, Maryland.

CHAPTER 7. FLUORIDES

HAROLD C. HODGE, Department of Pharmacology,
University of California School of Medicine
at San Francisco, San Francisco, California

FRANK A. SMITH, Department of Radiation Biology and
Biophysics, University of Rochester School of
Medicine and Dentistry, Rochester, New York

Fluorine is not a metal, and fluorides
do not create major contamination, but
the degree of interest shown in the
possibility of adverse effects is such
that the Planning Group decided to
include a short review in this volume.
(Ed.)

ECOLOGY

SOURCES AND CYCLES

Fluorine is a common element, 13th in abundance, constituting 0.077% of the earth's crust. Fluorspar (CaF_2) is the most important fluorine mineral of this ubiquitous element. Ground waters contain traces to comparatively large concentrations, 10 or more ppm. Plant and animal tissues contain traces or more; a few plants, e.g., dried tea leaves, as much as several hundred ppm, and some animal tissues notably bone and tooth mineral, up to several thousand ppm [1]. Rural air normally carries traces of F (fractions of ppb); in one study, city air particulates contained 0.01 to 3.9 μg F/m^3 [2]. Volcanoes sometimes release large amounts of fluoride. Industrial sources include the production of aluminum, fluorine, fluorocarbons, brick and pottery, steel and uranium; estimated emissions from some of these sources are given in Table 1.

TABLE 1 -- PRINCIPAL INDUSTRIAL SOURCES OF ATMOSPHERIC
FLUORIDES IN THE UNITED STATES

Industry	Approx. yearly atmospheric emission, tons
production of steel	4,600
brick, ceramics	18,500
aluminum	16,000
processing rock phosphate (fertilizers, H_3PO_4, P)	17,500
non-ferrous industries	4,000
burning of coal	4,100
	64,700

Personal communication, Office of Air
Programs, Environmental Protection Agency

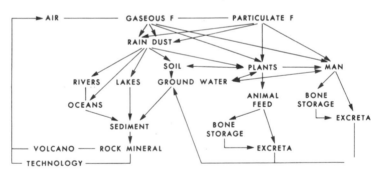

Fig. 1. Dispersion os fluoride in the biosphere.

Ecological significance appears to be limited to the immediate downwind sites of contamination but has not been given detailed, broadly-based study (Figure 1). Diets contribute about a milligram or less daily to the average man; drinking water accounts from traces to several milligrams daily (Tables 2, 3, 4). Fluorine-containing drugs contribute little or no F ion to the body (except certain compounds, such as the anesthetic agent, methoxyfluorane).

TABLE 2 -- DAILY DIETARY FLUORIDE (1)

Country	mg F in food	ppm F in water
Canada	0.18-0.3	0.1
England	0.3 -0.5	trace
Japan	0.47-2.66[a]	0.01-0.08[b]
Newfoundland	2.74 [c]	trace
Norway	0.22-0.31	0.01-0.07
Russia	0.6 -1.2	0.2 -0.3
Sweden	0.9	--
Switzerland	0.5[d]	--
U.S.A.	0.2 -0.3[d] 0.34-0.80	-- 0.1

a) Including 0.07-0.86 mg from green tea
b) mg F ingested
c) Including 1 mg from tea
d) Exclusive of that in drinking water

TABLE 3 -- FLUORIDE IN FOODS (1)

Food	Cholak, 1959	Recent data
	ppm F, fresh basis	
Meats	0.01-7.7	0.14-2
Fish	<0.10-24	0.05-19
Sardines		8-19
Shrimp		50[a]
Fish Meal		186
Citrus fruits	0.04-0.36	0.07-0.17
Non-citrus fruits	0.02-1.32	0.03-0.84
Cereals and cereal products	<0.10-20	0.18-2.8
Vegetables and tubers	0.10-3.0	0.02-0.9

a) Shrimp meat 0.4 ppm; shrimp shell 18-48 ppm.

165

TABLE 4 -- FLUORIDE IN BEVERAGES (1)

Beverage	F concn., ppm
Wine	0.0 -6.3
Beer	0.15-0.86
Tea infusion	0.1 -2.0
Instant (soln.)	0.2
Coffee bean	0.2 -1.6
Instant (powder)	1.7
Milk	0.04-0.55
Coca Cola	0.07
Orange juice	0.0 -0.05

AIRBORNE DISTRIBUTION

The literature contains reports of a number of industrial hygiene surveys in the fluoride-emitting industries, from which an appreciation of the hazard to man can be gleaned. Selected data from several of these surveys are shown in Table 5. The concentrations of airborne fluoride in individual samples range from traces (less than a few tenths mg/m^3) to maximal values of approximately 8 mg/m^3; averages range from 0.2 to 3.4 mg/m^3. Urinary fluoride excretions ranged from 1-5 mg/li. Increased irritation of the respiratory tract was reported in some instances, and an increased skeletal density to X-rays in some persons. No instances of significantly impaired health were noted. From industrial exposures, fluoride is absorbed across the respiratory membranes or moved by ciliary action into the gastrointestinal tract. Gaseous fluorides may penetrate the normal skin barrier in traces.

Table 6 lists several epidemiological studies where physical examinations have been made of children living near industrial plants emitting fluorides. For the most part, the effects are limited to increased enamel mottling and anemia. The data as reported make it well-nigh impossible to determine the intake of inhaled fluoride; in one instance, however (4), this was estimated to be approximately 20-30 per cent of the total daily intake.

TABLE 5 -- FLUORIDE IN INDUSTRIAL ATMOSPHERES (2)

Industry	F in Industrial Atmosphere		F in Urine		Comment
	range	mean	range	mean	
Aluminum	0.14-3.43 mg F/m³ (furnace rooms)	-	-	9.03 mg/24 hrs full-time male workers	25.4% incidence of x-ray abnormalities among 189 workers. 12.8% incidence of cough
				5.19 mg/24 hrs part-time male workers	
				3.64 mg/24 hrs full-time female workers	
	0.015-0.141 F/m³ (elsewhere in plant)	-	-	1.83 mg/24 hrs male 1.58 mg/24 hrs female	8.3% incidence of x-ray abnormalities among 60 workers. 6.9% incidence of cough
	0.033-0.048 mg F/m³ (control areas)	-	-	0.84 F/2mg 4 hrs	4% incidence of x-ray abnormalities among 75 persons
Phosphate/ fertilizer	1.78 -7.75 mg F/m³	3.38 mg F/m³	0.5-44.0 mg/11	5.18 mg/11	Minimal or questionable increased bone density to x-ray. Increased incidence albuminuria, pulmonary changes
	0.5 -8.32 mg F/m³	2.64 mg F/m³	0.2-43.0 mg F/11	4.51 mg F/11	No increase in bone density; increased incidence albuminuria, pulmonary changes
Magnesium foundries	-	0.143 mg F/m³	0.9- 4.1 mg F/11	-	Core shop workers
	-	0.286-0.714 mg F/m³	0.5- 7.5 mg F/11	-	Foundry workers
		0.314 mg F/m³ (near furnace)	1.0- 3.4 mg F/11	-	Furnace men. Probably 7 yr or more of exposure to produce x-ray changes
UF₆ production by gaseous diffusion process	0-24.7 ppm F	0.3-1.4 ppm F (avs. for different years)		1.1 mg F/11	Data over 8 years. No pulmonary x-ray changes attributable to respiratory irritant

167

TABLE 6 -- AIRBORNE FLUORIDE CONCENTRATIONS
IN THE VICINITY OF INDUSTRIAL FLUORIDE SOURCES

Industry	Country	Airborne Fluoride	Effects	Reference
Aluminum	Czechoslovakia	F concn. increased in dust, water, vegetation 1/4-1 ml from plant	145 children 6-14 yrs. old. Increases in erythrocytes, urinary F, mottled enamel. Decreases in hemoglobin, dental caries. Bone radiograms normal	(3)
	Czechoslovakia	0.02-0.14 mg/m^3	200 children 6-14 yrs. old. Increased erythrocytes, urinary F; decreased hemoglobin. Estd. daily F intake, 2.15 mg; of this, 0.4-0.7 mg estd. to come from air. Estd. intake of controls, 1.1 mg, none from air	(4)
	Russia	0.03-0.56 mg F/m^3 in villages near plants	Children 6-17 yrs. old. Increased incidence of mottled enamel and decreased incidence of dental caries	(5)
	Russia	Some samples 15x Russian MAC, at ranges of 500-1500 meters	449 children nursery to 12 yrs. old. Increased morbidity. Decreased hemoglobin, erythrocytes, blood sugar, phagocytic index	(6)
Phosphate	Russia	Concns. 16.2x Russian MAC at 500 meters, 3.3x at 3000 meters. Plus excessive SO_2, H_2SO_4, NO_x	1375 children living in villages 0.5-3 km from factory. Increased incidence of upper respiratory tract effects, tubercular and nontubercular changes in the lung	(7)

Analyses of the ambient atmospheres normally present over a number of U.S. cities have shown fluoride concentrations ranging between 0.01 µg/m^3 and 3.9 µg/m^3 (8). Fluoride absorption has not been measured, but assuming a volume of 20 cu. m. is inhaled per 24 hrs. with 100% retention, the maximal dose would be 0.004-0.04 mg daily. Martin (9) has thus calculated that an average man on a normal day in Central London would inhale 0.003 mg, with about a 10-fold increase on a day of thick fog and heavy pollution. In a heavily polluted area such as Stoke-on-Trent, the intake of the average man would be about 0.04 mg. Such amounts constitute negligible contributions to the total daily intake of fluoride from food and water (about 0.5-1.5 mg). These amounts of inhaled fluoride differ from those estimated by Balazova and Rippel (4) by an order of magnitude, a discrepancy which points up the difficulties of making reliable estimates of exposures from airborne fluoride.

TOXICOLOGY AND PHARMACOLOGY

MECHANISM OF ACTION

None of the many fluoride effects can be described adequately at a molecular level. From the numerous studies of fluoride actions in vitro and in vivo, plausible accounts of some of the factors probably responsible can be rendered. In excessive doses, "fluoride kills (in acute poisoning) by a blockade of the normal metabolism of the cells" (10). Enzyme inhibition by fluoride in vitro, especially enzymatic reactions involving divalent metal cations as cofactors, has been demonstrated for many kinds of enzyme systems (11). Fluoride concentrations shown to inhibit certain enzymes in vitro, e.g., about 10^{-3} M or 20 ppm, are attained in the blood (and presumably in the tissue fluids) of acutely and fatally poisoned rats and rabbits. "Vital functions, such as the origin and transmission of nerve impulses, cease" (10). Respiration fails before cardiac arrest (12). Bodily functions regulated by calcium, e.g., membrane transport, muscle contraction, blood clotting, are seriously deranged. "Cellular damage and necrosis produce massive impairment in the function of vital organs" (10). A shock-like syndrome precedes exitus, perhaps a result of a direct vasodilating action of fluoride (12).

In the skeletal structures and the teeth, fluoride in over-doses "mottles" bone by interference with the osteocyte and causes mottled enamel by impairing the work of the ameloblast. Evidence of stimulating effects on the osteocyte and on the ameloblast at optimal fluoride concentrations can also be deduced. New bone formation and exostoses have been repeatedly described presumably demonstrating increased osteocyte activity. By what means fluoride in excess produces calcification of ligaments in crippling fluorosis is a mystery. Whether fluoride can play a role in "nucleation" as calcification begins is not clear. Perfection of tooth development and form at lower doses may show the enhancement of ameloblast function.

The greater stability of the fluorapatite crystal lattice, and therefore the lesser solubility as contrasted with hydroxyapatite, presumably functions in the successful treatment of bone-destroying processes, e.g., senile osteoporosis, Paget's disease, and some bone cancers. The lower incidence of spontaneously fractured vertebrate in elderly persons who drink naturally fluoridated waters (3-6 ppm) can be attributed in part to the lower solubility and the decreased resorption of the larger, better-formed apatite tablets. The external-most layers of the dental enamel are higher in fluoride content when teeth calcify during exposure to fluoridated drinking water (natural or added fluoride). Lowered rates of dental caries can be explained in part by assuming a lower solubility of enamel mineral and in part by fluoride's inhibition of acid production in oral bacteria held on the tooth surface. Both the colorless organic plaque and the apatite mineral fix fluoride which thereafter may be released in low but significant concentrations when H ions from bacterial metabolism diffuse into and out of these structures.

Only recently has one of fluoride's principal metabolic effects been clearly demonstrated. In the liver of rats fed levels of fluoride sufficient to depress body growth, only citrate of many glucose metabolites is markedly altered in amounts: citrate concentrations increase several-fold (13). Another interesting metabolic action of fluoride (significance unevaluated) is its ability to stimulate adenyl cyclase activity in all tissues so far examined.

DISPOSITION

Absorption

With certain exceptions, inorganic fluoride is absorbed only as the fluoride ion (F^-). Fluoride salts vary considerably in solubility and therefore in the rate and extent of absorption following ingestion. Fluoride, absorbed in part from the stomach, is rapidly and efficiently taken up by the intestine. The rapidity of fluoride adsorption from the stomach and the intestine is striking; 90% of a small dose of NaF solution disappeared from the gastrointestinal tract of the rat in 90 minutes (14).

170

Distribution

Fluoride is carried by the blood and distributed through the body fluids and tissues in a fashion much like that of chloride; the fluoride distribution processes are somewhat slower. Only 10% of the F in human blood "normally" is exchangeable, i.e., fluoride ion, the balance appears to be bound to albumin; newly absorbed fluoride increased only the F^- ion concentration (15). Plasma exchangeable F concentrations are somewhat higher when drinking waters are fluoridated (0.04 ppm in plasma at 1 ppm, 0.01 ppm at 0.06 or less ppm) (16). Administered intravenously, F mixes promptly with the circulating fluid; thereafter the disappearance curve reveals at least two processes, one with a biological half-time of about 30 minutes, presumably bone deposition, the other with a half-time of 2 or 3 hours, presumably kidney excretion (10). Soft tissues characteristically exhibit levels of added fluoride below and paralleling blood concentrations (10). Soft tissues free from ectopic calcifications do not store fluoride.

Excretion

Urinary excretion ordinarily accounts for 90 - 95% of the fluoride excretion; the rest is mainly excreted in the feces (17). Sweating (if profuse) can account for up to 1/3 of the fluoride excretion. The rapidity of urinary excretion is remarkable; 20% of a 1 mg dose (as NaF) can be found in human urine 3 hours after drinking this minute amount in aqueous solution. No specific excretory mechanism is involved; tubular resorption is less efficient than for chloride. When the daily fluoride intake is suddenly increased, about half the increase appears in the urine during the ensuing 24 hours, the balance is deposited in the skeleton (Figure 2). When the increased dose is maintained day after day, the skeleton ultimately (months to years) becomes apparently saturated; a "steady state" develops such that nearly all the daily intake is accounted for in the urine (Table 7). Thus, urinary concentrations mirror nearly arithmetically drinking water concentrations (Figure 3). In renal failure, urinary excretion is understandably drastically impaired.

TABLE 7 -- RELATION BETWEEN FLUORIDE INGESTED IN FOOD AND WATER
AND THAT EXCRETED IN URINE AND FECES (1)

Age, yrs.	F in drinking water, ppm	Residence, yrs.[a]	Observation period, days	mg F IN fluid	food	sum	mg F OUT feces	urine	sum	Balance mg
33	< 0.1[b]	8	140	0.3	0.2	0.5	0.05	0.4	0.45	+0.05
35	2	10	96	2.4	1.2	3.6	0.4	2.9	3.3	+0.3
55	5.5	29	60	3.8	1.3	5.1	0.6	4.5	5.1	0
57	6.1	34	133	6.7	1.0	7.7	0.4	8.1	8.5	-0.8
57	8	19	140	11.3	2.5	13.8	1.4	10.4	11.8	+2.0
30	20	8	45	20.8	1.5	22.3	1.4	12.3	13.7	+8.6

a) Years in residence using the indicated water supply
b) Comparable data could not be found for individuals consuming water containing 1 ppm F.

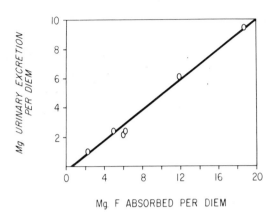

Fig. 2. Relation between absorption and urinary excretion of fluoride in man. (19)

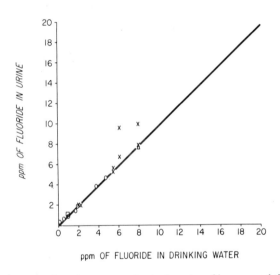

Fig. 3. Relation between fluoride concentration in the urine of humans and that in the water supplies used. (19)

Deposition

Fluoride, a prototype bone seeker (skeletal fluoride comprises 99% of the body burden), accumulates gradually during life even when intakes remain relatively constant (10). The levels ultimately reach "plateaus" during the 5th to 6th decade; the final concentration reflects the level of intake (Figure 4). The fluoride concentrations in tooth enamel and dentine also tend to be higher with increased intake. Part of the fluoride in bone mineral occupies elemental positions in the bone crystal lattice. The half-time for bone mineral as measured by fluoride turnover is about 8 to 10 years (Figure 5); the rest of the bone fluoride, located on crystal surfaces, is mobile, exchanging constantly with the fluoride ion in body fluids.

CLINICAL ASPECTS

In introducing the discussion of certain selected effects of fluoride, a few words of orientation are offered: 1. The emphasis will be placed on effects on man; comments about animals and about plants will be abbreviated. 2. Fluorides over a range of decreasing doses produce three recognized disease entities (acute fatalities, crippling fluorosis, and mottled enamel); at lower exposures, fluorides improve dental health (a major public health benefit) and a few other possible beneficial effects are under study. 3. Fluoride accumulates in bones and teeth, therefore in addition to dose, the duration of exposure is a controlling factor in the benefit or the kind and severity of injury. 4.

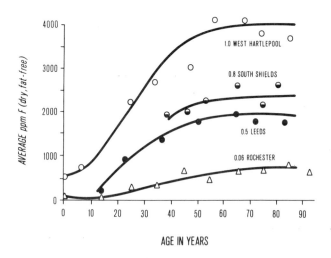

Fig. 4. Skeletal concentrations of fluoride in residents of West Hartlepool, South Shields, and Leeds England, and of Rochester, N. Y. (1)

174

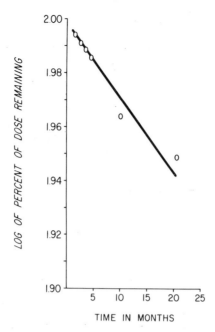

Fig. 5. Mobilization of skeletal fluoride in man. (19)

Attention will focus on a considerable number of carefully studied adverse fluoride effects. In addition, several non-specific "effects", e.g., lassitude, have been attributed to fluoride; these in general will not be discussed. 5. Fluoride is a metabolic poison, a complexer of important mineral ions, e.g., calcium or magnesium, which play key and versatile roles in cellular enzymatic processes and in membrane functions. Thus, fluoride effects are wide-spread and complicated; the mechanisms of these effects are not understood in general or in specifics.

EFFECTS OF SINGLE ORAL DOSES (Table 8)

Nearly 900 deaths in the U.S. between 1933 and 1966 were ascribed to fluoride poisoning (accidental, suicidal, homicidal). The course of the poisoning is characteristically rapid (2 to 4 hours) following "certainly lethal doses" by mouth of 2500-5000 mg soluble F (i.e., 5-10 g sodium fluoride) (10). Nausea, vomiting, abdominal cramps and diarrhea promptly develop; "severely poisoned individuals soon

TABLE 8 -- RESPONSES TO SODIUM FLUORIDE (10)

Effect	Dose mgF	Dose ppmF	Time	Comment
		SINGLE ORAL DOSE		
Death	2500-5000	--	2-4 hrs.	cellular metabolism blocked
nil	150-700	--	i.v. daily 9 days[a]	sensitive individuals may show minor response at lower doses
		REPEATED ORAL DOSE		
Kidney		100(w)[b]	mos.	dilated tubules; necrosis and regeneration in animals
Anemia	--	100 (f)	mos.	may be changed clotting time in animals
Reproduction	--	60 (f)	mos.	delayed oestrus; sterility; smaller less viable offspring of animals
Thyroid	--	50 (f,w)	yrs.	altered structure and/or function in animals
Body weight	--	40 (f)	5 yrs.	in dairy cattle calving and lactating; lower concentrations probably without effect
Crippling fluorosis	20-80 or more mg/day	--	yrs.	industrial disease in man; also in a few men of India and China; industrial or sometimes natural exposures in animals; osteosclerosis, sometimes osteoporosis
Treatment of osteoporosis	20-100 mg per day	--	mos., yrs.	promising. May cause epigastric distress or bone pain
No osteosclerosis	5 mg/day	4 (w)	yrs.	from industrial exposures or from community water supplies
Mottled enamel	--	2-8 or more (w)	1st 8 yrs.	pre-eruptive impairment of enamel-forming cells
Lessened osteoporosis	--	3-8 (w)	yrs.	single observation
Lessened aortic calcification?	--	3-6 (w)	yrs.	single observation
Decreased dental caries	--	0.4-1.5 (w)	lifetime	better formed, more resistant teeth
No adverse effects in man	--	0.2-1.2 (f)	lifetime	"normal" diets in several countries
No adverse effects in plants	--	1 ppb (0.001 ppm in air)		limit for susceptible plants

a) terminal cancer patient
b) w = water, f = food

collapsed exhibiting pallor, weak pulse, shallow respiration and cyanosis", coma preceded exitus. Whether the fluoride blockade of citrate utilization is partly responsible for death is unknown. The rapidity and efficiency of fluoride disappearance from body fluids is demonstrated by the improved survival after a few hours, and by the history of one patient who tolerated intravenous doses of several hundred milligrams of F daily for 9 days without adverse response.

EFFECTS OF REPEATED ORAL DOSES (Table 8)

KIDNEY -- Although dilated tubules have never been seen in man, they occur occasionally along with a shifting pattern of tubular necrosis and repair in rats receiving more than 100 ppm F in drinking water (some other species are more tolerant). Alterations in renal function reveal the severity of injury (10).

ANEMIA -- Whether the mild anemia identified in residents of India in fact resulted from the fluoride in the drinking waters containing 0 to 12 ppm F cannot be unequivocally concluded; little other evidence of adverse effect is available on man. Anemias and other hematologic changes have repeatedly been produced in animals given rations containing more than 100 ppm F (10).

REPRODUCTIVE PERFORMANCE -- Large enough daily doses of fluoride (over 60 ppm in the diets of animals) interfere in various ways with reproduction (10).

THYROID -- For almost a century, textbooks have repeated the story of a dog that developed a goiter from excessive doses of fluoride. Perhaps this reiteration together with the extraordinary ability of the thyroid to store iodine, a sister halogen, explains the apprehension often detectable that fluoride may injure the thyroid. Many studies of animals serve to establish the fact that repeated doses, for the most part exceeding 50 ppm F in feed or water, do indeed alter thyroid structure or function (10). Reassurance can be drawn from the inability of the thyroid to store fluoride. In man, fluoride (up to 3 ppm in drinking water) neither blocks nor enhances goiter development in areas deficient in iodine. Furthermore, fluoride in this concentration range shows no interference with the prophylactic effect of iodine (10).

BODY WEIGHT; GROWTH -- The animal most susceptible to weight loss from F feeding is the dairy cow in the cycle of calf-bearing and lactation;

forage containing 40 ppm F will in 4 1/2 to 5 years
induce an "unthrifty" condition with gradual loss of
weight (10). Weight loss from fluoride has never
been seen in man, nor has growth retardation (if a
purported stunting in a primitive Japanese tribe can
be discounted). Certainly in the U.S., growth of
children where the community water supply was
fluoridated at 1 ppm, matched the growth of
comparable children in a control unfluoridated city
nearby. The best and longest study of normal human
growth ever made vouches for this conclusion (10).
Two reports of teratogenic effects of fluorides are
available; doses of more than 12-24 mg/kg given
parenterally or orally to pregnant mice and rats
produced abnormalities of liver, kidney, skull, jaw
bones, and teeth (18,19).

SPECIAL PROBLEMS

CRIPPLING FLUOROSIS -- Described forty years ago
as an industrial disease of cryolite workers, the
symptom complex has been recognized in a few small
groups of native workmen in China, in India, in the
Middle East, and in South America (10) as well as in
animals consuming vegetation or drinking water
excessively contaminated with fluorides.
Hypermineralization of the skeleton (sometimes with
"motheaten" loci of hypomineralization),
calcification of ligaments, and bony outgrowths
(exostoses) form a diagnostic triad; stiff, painful
joints and immobilization (e.g., of parts of the
spine) explain the crippling effects. In man,
prolonged intake, as by inhalation, of 20 to 80 mg F
or more daily, over a period of years is required.

CAN FLUORIDE PREVENT OR REVERSE OSTEOPOROSIS?
-- Evidence drawn from a score of studies during the
past decade offers hope that some regimen of fluoride
therapy may reduce the morbidity of this most
widespread and crippling of the diseases of the
elderly, a disease without specific therapy. The
reported responses are by no means uniform. For
some, fluoride therapy must be discontinued because
epigastric distress or bone pain become intolerable.
Others tolerate 20-100 mg F per day for months in
some cases with signs of increased mineral mass
(osteosclerosis), positive calcium balance, or
reduction in bone pain. The safety of fluoride

therapy has not yet been fully substantiated; current tests center on this problem (20).

FLUORIDE INTAKE NOT INDUCING OSTEOSCLEROSIS -- The pelvic vertebrae first show osteosclerosis in individuals with high fluoride intake. Osteosclerosis has not been found: a) when industrial exposures are controlled such that urinary concentrations remain below 5 mg/li, or b) in communities where the drinking water contains 4 ppm or less (10).

MOTTLED ENAMEL -- During the first eight years of life, when the deciduous and later the permanent dentition are forming (pre-eruption), excessive fluoride intake (2 to 8 ppm or more in community water supplies) produces a disfiguring malformation of the enamel surface which becomes stained on contact with saliva. Minor white spots with no esthetic damage can be found in the teeth of persons drinking waters containing traces (less than 0.1 ppm F) and up to 1 or 1.5 ppm in temperate climates. In hot climates, lower F concentrations are associated with brown stain, e.g., at 0.4 to 0.5 ppm in southern Japan. More children suffer more severe mottling as fluoride concentration increases (Figure 6); however, some children retain normal surfaces even at 8 ppm F (10).

LESSENED OSTEOPOROSIS AND AORTIC CALCIFICATION IN THE ELDERLY -- A single report of a survey of osteoporosis (sponstaneously fractured vertebrae) and of the frequency of aortic calcification in elderly residents of two areas of North Dakota, one with "high" F drinking waters (3 to 6 ppm) and the other with "low" F drinking waters (0.3 - 0.5 ppm), showed less osteoporosis in the "high" area especially in women 60 years old and older, and less aortic calcification upon radiographic examination in both men and women (21).

DECREASED DENTAL CARIES -- About 30 years ago, careful epidemic studies revealed to a doubting dental world that low concentrations (0.5 - 1.5 ppm) of fluoride in community water supplies were associated with fewer and smaller, more slowly progressing cavities (Figure 6). Twenty-five years ago the first deliberate additions of fluoride brought water supplies low in fluoride (less than 0.1 ppm) up to an "optimal" concentration for temperate climates of about 1 ppm. A lower fluoride concentration (<1 ppm) is recommended for communities in warmer climates, e.g., 0.7 ppm for a mean annual temperature of 70 F. Today no serious scientific question of fluoride's efficacy remains, the benefit -- 1/2 to 1/3 as many cavities in childhood -- persists into adult life of residents. Safety is supported by a mass of evidence; no proven injury of young or old, sick or healthy has occurred. Large-scale, controlled epidemiologic studies of populations drinking fluoridated vs. fluoride-low water supplies have not been conducted; conclusions are based in part on clinical opinions. Sophisticated evaluations of the risks of two groups of patients are needed: a) patients with advanced renal disease, especially those with abnormally high water intake, and b) patients possibly allergic to fluoride.

NO ADVERSE EFFECTS -- Without proof, it may be assumed that the daily intake of 0.2 - 1.2 mg F is "normal"; this range includes estimates of the dietary fluoride content in several countries.

ESSENTIALITY? -- The ubiquitous nature of fluoride has blocked all attempts to prepare a fluoride-free ration. Thus, the simplest approach to

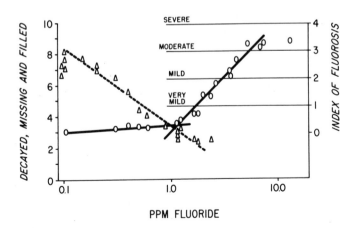

Fig. 6. Relations among the incidence of decayed, missing and filled teeth, severity of dental fluorosis and log fluoride concentration in water (22)

the question of an essential role is denied. Attempts to purify dietary components have been only partially successful (diets containing 0.007 ppm F have been used), and animals consuming such diets showed no ill effects (23).

EFFECTS OF INHALED FLUORIDES -- Inhaled fluoride dusts and gases (e.g., hydrogen fluoride, HF) furnish more or less fluoride depending presumably on dustiness, solubility, particle size, and other factors. Elemental fluorine and hydrogen fluoride in high concentrations may be fatally irritant to the lung. The Emergency Exposure Limits for single exposures without disability are low, e.g., for 10 minutes, 15 ppm in air of fluorine (F), 20 ppm in air of HF. The Threshold Limit Values (TLV) for the working week are 0.1 ppm for F , 3 ppm for HF. Fluoride dusts in general have a TLV of 2.5 mg/m^3 of air.

RECOGNITION AND CONTROL OF F EXPOSURES

Only traces of fluoride (less than 0.05 μg F/m^3) were found in 88% of more than 9000 samples of suspended water soluble particulates obtained by the U.S. Air Sampling Network at urban sites; over 98% of more than 2000 non-urban samples were below this limit (24). Breathing such air, a man would inhale less than 1 μg F in 24 hours, a negligible amount. Relatively large quantities of fluorides emitted from volcanoes, blown as dust from out-cropping phosphate rock, or released in industrial operations, such as superphosphate plants and aluminum production, have injured vegetation, animals grazing nearby, and rarely (if ever) neighborhood residents.

VEGETATION -- Some species and varieties of plants are highly sensitive to gaseous fluorides, others are resistant. Sensitive plants develop chlorosis and necrosis (generally marginal leaf burn); these visible effects can be used as indicators of gaseous fluoride exposure. Growth or yield are reduced in certain other plants (Figure 7). Particulates may be much less toxic depending on the solubility and also on particle size (>0.5μ , relatively ineffective). Man inhaling ambient air containing less than 0.05 μg of soluble F per m^3 and retaining all of it would absorb negligible amounts,

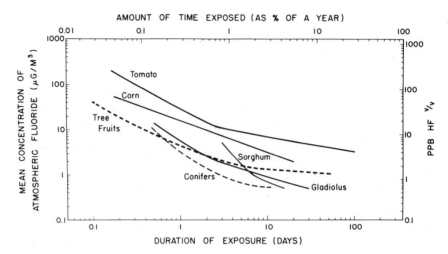

Fig. 7. Response of plants to hydrogen fluoride. (25)

and even with air concentrations of 3.5 ppb would absorb only a fraction of a milligram daily.

ANIMALS -- Grazing animals receive fluoride principally from fluoride dust on the foliage. High doses cripple; animals become lame, unable to graze except on their knees. Lower but toxic doses impair tooth formation, teeth are mottled and brittle and wear away to useless stumps. "Unthriftiness" and slowly developing weight loss mark the borderline toxicity when the forage of dairy cows contains more than 40 ppm F (dry weight basis); 30 ppm F can be taken as a "no effect" level.

MAN -- Food and water contaminated by fluoride dust windblown from phosphate outcroppings were held responsible for mottled enamel and bone changes in residents of North Africa. Industrial effluents in Scotland and in the United States and Canada have in some studies given no evidence of effect on human health. Two reports claim that illness resulted in nearby residents of an aluminum plant, fertilizer factories, and an iron foundry; neither report is convincing. One describes a "subacute fluorosis", never previously observed, comprising among other signs and symptoms, hepatic disorders, thyroid enlargement, stiff back, gastric symptoms, nausea, and cough. The other equates this "non-skeletal phase of fluorosis" with a newly coined "neighborhood" fluorosis with symptoms relating to the musculoskeletal, respiratory and gastrointestinal systems plus chizzola maculae. Gradually increasing general malaise and exhaustion leading to complete disability characterize the disease, also never previously observed. Four other reports in other countries, two of these in Russia, ascribe muscle and joint pains, anemia, increased lung disease, mottled enamel and reduced dental caries (in children) to fluoride effluents. Data supporting the causal relation between illness and fluoride exposure are meager at best and unsatisfactory (2).

INDUSTRIAL EXPOSURES -- Burns, thermal from F_2 and chemical from HF, are well-established hazards. Soluble fluorides produce contact dermatitis. Irritation of the upper respiratory tract, eyes, sometimes of the lungs (cough) are frequent complaints. Gastrointestinal symptoms are occasionally reported. With long-continued, excessive exposures osteosclerosis can be shown radiographically. Crippling fluorosis as described by Roholm has rarely if ever been seen since. Air sampling forms the basis for good industrial hygiene control; urine analyses give reliable evidence of the effectiveness of the protective program.

WATER-BORNE FLUORIDE -- A few cases of crippling fluorosis in India and China have been attributed to fluoride in surface well waters in concentrations of less than about 10 ppm (10). The victims may have been chronically malnourished, overworked under conditions of continuing severe heat stress, sufferers from intercurrent disease; their fluoride intake was unknown. Mottled enamel (brown stain) has been reported from many parts of the globe, in the temperate regions where drinking waters contain more than 2 ppm F, in hot regions at much lower F concentrations (0.4 to 0.7 ppm). Defluoridation can be achieved by a calcium phosphate treatment.

ANALYTICAL METHODS

Analysis for fluorine and fluorides has been extensively reviewed in two recent monographs (26,27). Additional appropriate references for measuring fluoride in blood

(28), urine (29,30), and air (31) are
appended here.

REFERENCES

1. Hodge, H. C. and Smith, F. A. (1970).
    Minerals: Fluorine and dental caries. In: R.
    F. Gould (ed.) Dietary chemicals vs. dental
    caries. Adv. in Chem. Series 94, Am. Chem.
    Soc., Washington, D. C.

2. Hodge, H. C. and Smith, F. A. (1970). Air
    quality criteria for the effects of fluorides on
    man. J. Air Pollut. Cont. Assoc. 20: 226.

3. Macuch, P. et al. (1963). Hygienic analysis
    of the influence of noxious factors on the
    environment and state of health of the
    population in the vicinity of an aluminum plant.
    J. Hyg. Epidemiol. Microbiol. Immunol. 7:
    389.

4. Balazova, G. and Rippel, A. (1968).
    Investigation of the health conditions of people
    living in the neighborhood of an aluminum
    factory. Med. Lavoro. 59: 376.

5. Sadilova, M. S. (1957). Atmospheric air
    pollution with fluorine as the cause of
    fluorosis in children of inhabited localities.
    In: B. S. Levine. Limits of Allowable
    Concentrations of Air Pollutants. Book 3, U.
    S. Dept. Commerce, Washington, D. C.

6. Kvartokvina, L. K. et al. (1968). Effect of
    discharges from an aluminum works on health of
    children. Hyg. Sanit. 33: 106.

7. Lindberg, Z. Ya. (1960). Fluorides and dental
    caries in children living near waste from a
    superphosphate plant. Gig. Sanit. 25: 89.

8. U. S. Dept. Health, Education, and Welfare.
    (1958). Air pollution measurements of the
    National Air Sampling Network. Publication No.
    637. Washington, D. C.

9. Martin, A. E. (1970). Fluorides and general health In: Y. Ericsson (ed.) Fluorides and Human Health, WHO Monog. 59, World Health Organization, Geneva, Switzerland.

10. Hodge, H. C. and Smith, F. A. (1965). Biological effects of inorganic fluorides. In: J. H. Simon (ed.) Fluorine Chemistry vol. 4, Academic Press, New York.

11. Wiseman, A. (1970). Effect of inorganic fluoride on enzymes. In: F. A. Smith (ed.) Pharmacology of Fluorides, Part 2, Springer-Verlag, New York.

12. Caruso, F. S., Maynard, E. A. and DiStefano, V. (1970). Pharmacology of sodium fluoride. In: F. A. Smith (ed.) Pharmacology of Fluorides, Part 2, Springer-Verlag, New York, Inc., New York.

13. Shearer, T. R. and Suttie, J. W. (1970). Effect of fluoride on glycolytic and citric acid metabolites in the rat liver. J. Nutrit. 100: 749.

14. Zipkin, I. and Likins, R. C. (1957). Absorption of various fluorine compounds from the gastrointestinal tract of the rat. Am. J. Physiol. 191: 549.

15. Taves, D. R. (1968). Evidence that there are two forms of fluoride in human serum. Nature 217: 1050.

16. Smith, F. A., Gardner, D. E. and Hodge, H. C. (1950). Fluoride content of blood and urine as a function of the fluorine content in drinking water. J. Dent. Res. 29: 596.

17. Hodge, H. C., Smith, F. A. and Gedalia, I. (1970). Excretion of fluorides. In: Y. Ericsson (ed.) Fluorides and Human Health. WHO Monog. 59, World Health Organization, Geneva, Switzerland.

18. Fleming, H. S. and Greenfield, V. S. (1954). Changes in the teeth and jaws of neonatal Webster mice after administration of NaF and

CaF_2 to the female parent during gestation. J. Dent. Res. 33: 780.

19. D'Angelo, M. and Esposito, U. (1965). Histological observations on the newborn of rats intoxicated with sodium fluoride. Ann. Stomatol. 14: 835.

20. Hodge, H. C. and Smith, F. A. (1968). Fluorides and man. Ann. Rev. Pharmacol. 8: 395.

21. Bernstein, D. S. et al. (1966). Prevalence of osteoporosis in high- and low-fluoride areas in North Dakota. J. Am. Med. Assoc. 198: 499.

22. Hodge, H. C. and Smith, F. A. (1956). Some public health aspects of water fluoridation. In: J. H. Shaw (ed.) Fluoridation as a Public Health Measure. Am. Assoc. Adv. Sci., Washington, D. C.

23. Maurer, R. L. and Day, H. G. (1957). The non-essentiality of fluorine in nutrition. J. Nutrit. 62: 561.

24. Thompson, R. J., McMullen, T. B. and Morgan, G. B. (1970). Fluoride concentrations in the ambient air. Presented at 63rd Ann. Mtg., Air Pollut. Control Assoc.

25. McCune, D. C. (1969). On the establishment of Air Quality Criteria, with reference to the effect of atmospheric fluorine on vegetation. Air Quality Monog. 69-3, Am. Petroleum Ist., New York.

26. Elving, P. J., Horton, C. A. and Willard, H. H. (1954). Analytical chemistry of fluorine and fluorine-containing compounds. In: J. H. Simon (ed.) Fluorine Chemistry Vol. 2, Academic Press, New York.

27. MacDonald, A. M. G. (1970). Pharmacology of fluorides. In: F. A. Smith (ed.) Handbook of Experimental Pharmacology, vol. 20, Springer-Verlag, New York, Inc., New York.

28. Taves, D. R. (1968). Determination of submicromolar concentrations of fluoride in biological samples. Talanta 15: 1015.

29. Sun, Mu Wan (1969). Fluoride ion activity electrode for determination of urinary fluoride. Am. Ind. Hyg. Assoc. J. 30: 133.

30. Neefus, J. D., Cholak, J. and Saltzman, R. E. (1970). The determination of fluoride in urine, using a fluoride-specific ion electrode. Am. Ind. Hyg. Assoc. J. 31: 96.

31. American Industrial Hygiene Association. (1970). Analytical guide: Fluorine and its Compounds. Am. Ind. Hyg. Assoc. J. 31: 397.

PART III

ADDITIONAL CONSIDERATIONS

CHAPTER 8. NUTRITIONAL ASPECTS OF METALS

M. R. SPIVEY FOX, Division of Nutrition,
Food and Drug Administration, Department of
Health, Education, and Welfare, Washington, D. C.

Not all the effects of metals in biological systems are adverse to the health of the organism. Several metals have long been recognized as essential for the life processes of microorganisms, plants, and animals, including man. A number of non-metallic inorganic elements perform biological functions similar to those of the metals. In the strictest sense, only those elements can be designated essential that are specifically and irreplaceably required for some direct metabolic effect that is mandatory for survival of the species through succeeding generations. Beneficial health effects, which are unrelated to survival, have been shown for other elements. For simplicity in this review, all inorganic elements required for either essential or beneficial effects will be treated as essential elements.

Consideration of the relationships between nutrition and response to toxic elements raises two important questions. Is an element toxic due to interference with the function of a nutritionally essential element? Can the degree of response to the toxic element be modified by diet, including the intake of nutrients other than the inorganic essentials? Examples of such relationships may be found in the earlier chapters on the individual metals. The present purpose is to review briefly the elements that are essential, the general mechanisms by which they function, and the physiological processes that they affect.

THE ESSENTIAL ELEMENTS

The importance of the macro-elements sodium, potassium, chlorine (as chloride), calcium, phosphorus, and magnesium has been recognized for many years. In general, population exposure to these elements is not a health hazard. The normal metabolism of the essential elements may be disturbed by toxic elements and the dietary intake of these elements may greatly affect the dose-response relationship of the toxic elements. For example, it has been shown that the toxic effects and accumulation of lead are increased many-fold by inadequate intake of calcium (1). Since information on the macro-elements is readily available in any standard textbook of biochemistry, they will not be further considered here.

In the past few years research on trace elements has been greatly facilitated and stimulated by improvements in methods of analysis and development of techniques for producing ultraclean environments and highly purified diets (2). The trace elements, which are required by man in microgram to milligram quantities per day, can pose problems with respect to obtaining both adequate and toxic levels of intake. The concentrations of some trace elements in food may vary widely as a function of the level available to the original plant or animal. Trace elements in soil, fertilizer, and water vary over a wide range. In addition, many elements enter the body through the lungs, from either particulate matter or fumes.

The trace elements required in the largest amounts are iron, zinc, copper and manganese. Their essentiality has been recognized for several decades. Requirements have been shown for lesser quantities of selenium, chromium, cobalt, iodine, fluorine (as fluoride), and molybdenum. Quite recently evidence has been presented that tin, nickel, and vanadium may perform essential functions in animals receiving diets deficient in these elements.

BIOLOGICAL FUNCTIONS OF ESSENTIAL TRACE ELEMENTS

Detailed consideration of the functions of all essential trace elements is far beyond the scope of this review. Many, such as those for iron and iodine, have been generally well known for years. The focus here will be on more recent developments. Further, the function of the trace elements will be considered in relation to environmental toxicants, chelating agents in the food supply, and common nutrient deficiencies in the United States and other countries. Again, examples will be chosen to illustrate a point rather than to provide a comprehensive survey. Regrettably space does not permit citation of all original references; however, reviews provide ready access to this literature for trace elements in general (3,4), as well as for copper (5,6), zinc (7-10), manganese (11,12), fluoride (13,14), selenium (15,16), and chromium (17,18).

Table 1 summarizes the effects of trace element deficiencies in man and animals. Table 2 shows very general estimates of requirements. The intakes of normal population groups can serve as a very crude approximation of need in the absence of more accurate information. Together with data from precise animal experiments, it is possible to make reasonable estimates of requirement and factors that will influence availability of the element and the requirement of man.

Iron

This element is an essential component of myoglobin, hemoglobin, and many oxidation-reduction enzymes. Iron-responsive anemia is common throughout the world, particularly in women and young children (19). The availability of different forms of iron suitable for fortification of foods is presently receiving considerable attention. Low intakes of folic acid, ascorbic acid and protein are also characteristic of many population groups. Deficiencies of these nutrients may be contributing factors to the observed anemia, particularly those of ascorbic acid and protein since they can increase the availability of iron.

It seems reasonable to expect that a person receiving an inadequate dietary intake of one or more of these nutrients might be more susceptible to the anemia-producing effects of toxic elements. Conversely, there might be protection against the

TABLE 1 -- EFFECTS OF TRACE ELEMENT DEFICIENCIES IN MAN AND ANIMALS[a]

Element	Species	Defect
Iron	Man	Anemia
	Rat	Anemia
	Chick	Anemia, achromotrichia in certain breeds
Copper	Man	Anemia (rare), hypogeusia
	Rat	Anemia, retarded growth, achromotrichia
	Sheep	Brittle bones, neonatal enzootic ataxia associated with demyelination and degeneration of the nervous system, achromotrichia, defective wool structure
	Chick	Anemia, dissecting aneurysms, ataxia, achromotrichia in certain breeds
Zinc	Man	Hypogonadism, short stature, hypogeusia
	Rat	Death, retarded growth, hypogonadism, skin and esophageal lesions, alopecia, embryonic deformities
	Chick	Death, retarded growth, shortened and thickened long bones, defective feathering, embryonic deformities
Manganese	Man	None demonstrated
	Rat	Retarded growth, defective bone mineralization, testicular degeneration, defective ovulation, congenital paralysis and incoordination due to deformed or absent otoliths
	Chick	Perosis, shortened and thickened long bones, decreased egg production and shell strength, lowered hatchability, embryonic chondrodystrophy
Cobalt	Man	None, apart from vitamin B_{12}
	Animals	None, apart from vitamin B_{12}
Molybdenum	Man	None demonstrated
	Rat	None demonstrated
Fluoride	Man	Dental caries
	Rat	Dental caries
Iodine	Man	Goiter
	Rat	Goiter
	Chick	Goiter
Selenium[b]	Man	Anemia
	Rat	Liver necrosis
	Lamb	White muscle disease
	Chick	Exudative diathesis, muscular dystrophy
Chromium	Man	Impaired utilization of glucose
	Monkey	Impaired utilization of glucose
	Rat	Impaired utilization of glucose, corneal opacity[c]

a) For sources of original literature, see reviews and text.
b) Except for pancreatic atrophy, all defects depend upon concomitant deficiencies of vitamin E and selenium.
c) Occurs only in animals fed diets low in both chromium and protein.

TABLE 2 -- RANGES OF REQUIREMENTS AND/OR TYPICAL DAILY INTAKES OF ESSENTIAL TRACE ELEMENTS BY HUMAN BEINGS*.

Element	Amount per day
	mg
Iron	10 - 18
Zinc	6 - 15
Manganese	1.25 - 6.5
Copper	1.0 - 3.8
Fluoride	0.5 - 1.7
Iodine	0.10 - 0.14
Molybdenum	0.10
Chromium (III)	0.06 - 0.36
Selenium	0.03 - 0.05
Cobalt	0.015- 0.160

*Ranges are to show general magnitude of need. For the few specific recommendations, see reference (4).

toxic elements by a diet containing amounts of these nutrients in excess of the levels generally considered adequate. We have found that Japanese quail (Coturnix coturnix japonica) receiving supplemental ascorbic acid were markedly protected against the anemia produced by cadmium (20).

Copper

A deficiency of copper also results in an anemia. The copper intake of the U. S. population has been considered to be adequate, with the possible exception of infants maintained exclusively on a milk diet. Recently, however, analysis of school lunches in five regions of the U. S. showed that the copper content was low (21). The effects of the anemia-producing toxic elements might be more severe in individuals with inadequate copper intake. Copper, zinc, and cadmium have been found to bind to the same protein in the duodenal mucosa of the chick (22). This protein is thought to function in the intestinal absorption and/or transport of these three elements.

Copper has been shown to have a functional role in elastin synthesis as reviewed by Carnes et al. (23). There is evidence that a copper-containing enzyme is responsible for the oxidative deamination of the epsilon-amino group of lysine to produce the elastin cross-links desmosine and isodesmosine. In copper-deficient animals the arterial elastin is more soluble and there are decreased concentrations of desmosine and isodesmosine. The inadequate elastin results in defects in vascular integrity, including dissecting aneurysms, in copper-deficient animals.

d-penicillamine (β,β-dimethylcysteine) chelates copper very avidly and has been used therapeutically to remove excess copper stores from patients with Wilson's disease (24). In these and other patients receiving penicillamine, a loss in the sense of taste was observed. This function was restored when copper sulfate was administered (25).

Zinc

A deficiency of zinc has been produced in many animals. In 1955 an outbreak of parakeratosis in commercial swine was traced to inadequate zinc intake (26). Young men with a syndrome of dwarfism, hepatosplenomegaly and hypogonadism, which is common in Mid-Eastern countries, responded dramatically to the combination of a nutritionally adequate diet and a large daily supplement of zinc (27). Dwarfism has now been studied in two females and there was a similar response to zinc and an adequate diet (28). The nutritional dwarfism is thought to be an extreme form of generally widespread zinc deficiency occurring in malnourished villagers (29).

Phytic acid, the hexaphosphate ester of inositol that is present in all seeds, can bind zinc and reduce its intestinal absorption. The zinc deficiency in the dwarfs has been attributed to the high intake of phytic acid present in unleavened whole wheat bread, their principal dietary staple. Calcium, particularly in the presence of phytic acid, can also reduce the absorption of zinc. Synthetic chelating agents can either increase or decrease the bioavailability of zinc depending on how tightly the chemical binds zinc. Ethylenediamine tetraacetic acid (EDTA), a common food additive, increases the absorption and availability of zinc in diets containing phytic acid.

Geophagia, the practice of eating clay or dirt, occurs throughout the world (30). All of the dwarfs studied in Iran ate clay; however, it is not clear what specific effect this had on their zinc deficiency (31). Not only could clays affect mineral absorption, depending on their own particular composition, but they could also serve as a source of toxic elements.

Zinc seems to be unique among trace elements with respect to body stores. When zinc is removed from the diet, the plasma zinc concentration falls to the deficiency range within 24 hours. Hurley has found that normal pregnant rats bore small, malformed zinc-deficient pups if zinc was removed from the diet of the mother in the early stages of pregnancy (32). It appears that either zinc stores are very small or non-functional zinc is so tightly bound that it cannot be shifted within the body to meet a crucial need. A normal animal that is otherwise well nourished and protected from stresses does not appear to be immediately affected adversely by removal of dietary zinc. Conversely, many of the tissues in a severely zinc-deficient animal have normal concentrations of zinc. The factors that regulate movement of zinc within the body are poorly understood, although the adrenal cortex can have a marked effect (33).

In addition to nutritional dwarfism, large supplements of zinc have been reported beneficial in wound healing (34), circulatory problems associated with atherosclerosis (35), and iodiopathic hypogeusia (33), a loss of the sense of taste. These studies suggest that a number of factors, as yet uncharacterized, may affect zinc requirement.

Manganese

There is clearcut evidence from animal experiments that manganese is an essential element, and as a result it is considered so for man. The most characteristic defects in manganese deficiency involve bone formation. High levels of dietary manganese interfere with iron absorption. Manganese is widely distributed in plant and animal foods, so that a deficiency in man has not been recognized. A preliminary report indicated that a higher incidence of human cancer in Finland was geographically related to soils low in easily soluble manganese (36).

Cobalt

A serious "wasting disease" in cattle and sheep grazing in a western section of Australia was found to be due to a deficiency of cobalt in the soil and forage. It was subsequently shown that vitamin B_{12} could replace cobalt. There is no evidence that cobalt has any essential functions other than as a component of vitamin B_{12}. High levels of cobalt will produce polycythemia but there is no evidence that this mechanism is involved in the hematopoietic function of vitamin B_{12}. The only source of vitamin B_{12} in foods derives from microbiological synthesis. Severe cardiac failure has been associated with consumption of large quantities of beer containing cobalt added for purposes of foam stabilization (37). Some of these beers contained a total of only 1.2 - 1.5 ppm.

Molybdenum

The activity of the enzyme xanthine oxidase is dependent on molybdenum. In animals fed diets low in molybdenum the level of this enzyme was decreased in the intestine and liver but no adverse physiological effects were observed. Addition of tungsten to the diet of chicks caused marked excretion of molybdenum, lowered capacity to oxidize xanthine, poor growth and high mortality. Supplements of molybdenum corrected these adverse effects. Because of these experiments and the importance of xanthine oxidase, molybdenum is considered to be essential for man, probably in very small amounts.

Complex interrelationships exist between molybdenum, copper, and sulfate. Molybdenum toxicity is markedly reduced by copper sulfate. Copper poisoning has been reported with moderate copper intakes when dietary molybdenum and sulfate were low. Copper deficiency, including low tissue copper levels, has been observed when both molybdenum and sulfate intakes were high. It is not known whether comparable relationships exist in man.

Fluoride

It has been well established that consumption of drinking water containing 1 ppm fluoride during childhood will protect against dental caries. Very slight mottling of the dental enamel may occur in a small percentage of children in temperate climates drinking water containing 1-1.5 ppm fluoride. This degree of mottling is not aesthetically damaging since there is no irregularity of the enamel surface and no brown stain.

A survey of persons 45 years of age and older was conducted in areas of North Dakota where fluoride in the water ranged from 4.0 to 5.8 ppm and in other areas where the fluoride content ranged from 0.15 to 0.30 ppm (13). In comparison to persons drinking low-fluoride water, the population groups receiving the high fluoride had bones of greater density, fewer collapsed vertebrae in women and less calcification of the abdominal aorta in the men. These benefits to the older persons were related to a fluoride content of water that causes mottling of the enamel in children. Further studies are needed to evaluate the effectiveness of lower levels of fluoride on these responses. Of all the essential trace elements, fluoride has the narrowest margin between beneficial and undesirable consequences. No public health measures have yet been devised whereby the appropriate beneficial levels of fluoride could be given to the adults without causing the undesirable mottling of teeth of the children.

Several workers have tried unsuccessfully to show that fluorine is an essential element in the strictest sense. Most recently a diet containing 0.005 ppm fluoride was prepared from sorghum and soybeans grown hydroponically (38). The fluoride-low and supplemented diets were fed from weaning to 91 days of age. The only difference observed in rats receiving the fluoride-deficient diet was that isocitric dehydrogenase was elevated in the serum and decreased in the liver.

Iodine

Dietary iodine is required for formation of several iodinated compounds in the thyroid. The most important of these are thyroxine and triiodothyronine, two hormones that affect all tissues of the body in a variety of ways. A deficiency of iodine results in goiter, which is preventable by the use of iodized salt.

Selenium

The distribution of selenium in the soil ranges from very low to high levels. This geographical distribution pattern results in areas where domestic animals develop symptoms of deficiency and others where animals are poisoned by the excess selenium in plants and water. Man may also be affected in the high selenium areas. Foods containing 5 ppm selenium are considered hazardous.

As little as 0.02-0.1 ppm selenium from inorganic salts can protect against the diverse manifestations of deficiency such as liver necrosis in rats, exudative diathesis in chicks and turkeys and white muscle disease in lambs. A level of 0.007 ppm selenium in the organic form of Factor 3 (as yet uncharacterized) is active against liver necrosis. In all of these disorders both selenium and vitamin E must be deficient for the syndrome to occur. Pancreatic atrophy and malfunction have been reported in chicks due to selenium deficiency alone (39).

There is little evidence that selenium deficiency occurs in man. Malnourished, anemic Jordanian infants showed a reticulocyte response to 30-50 ug elemental selenium per day (40). Other limited data indicate that selenium deficiency may be a contributory factor to kwashiorkor in infants.

The toxicity of selenium may be modified by diet. Feeding either linseed oil meal (41) or a combination of methionine and vitamin E (42) greatly reduced the liver damage caused by a toxic intake of selenium.

Chromium

An excellent comprehensive review of chromium and factors related to its biological functions has been written by Mertz (18). The occurrence of impaired glucose tolerance responsive to chromium (III) was first reported by Mertz and Schwarz in

1955. This response has now been observed in a wide range of experimental conditions as well as in human beings. From in vitro tests, the effect of chromium on glucose utilization depends on the presence of insulin. Usually only the trivalent form of chromium has had biological activity. Chromium (III) that has been incorporated into a small organic compound by yeast is much more active in in vitro systems than inorganic chromium. Only the organically-bound form of chromium from yeast was transferred across the placenta of rats when compared with inorganic chromium and a variety of chromium complexes.

The fetus receives a significant store of chromium. Tissue analysis shows that there is a gradual trend toward chromium depletion throughout life. Impaired glucose tolerance responsive to chromium has been observed in middle-aged and elderly individuals. The chromium content of hair appears to be closely correlated with chromium status in human beings (43).

"UNIDENTIFIED" TRACE ELEMENTS

By use of ultra-clean environments it has been possible recently to show beneficial effects in experimental animals of three "new" trace elements. Schwarz et al., used a highly purified amino acid diet to produce growth retardation in rats (44). Supplementation with 1 mg of tin per kg of diet produced a significant growth effect. Growth was increased 59% by 2 mg tin (from stannic sulfate) per kilogram of diet. These levels of tin are comparable to amounts in foods. The data are strongly indicative of an essential role of tin for animals.

In other experiments, dietary ingredients that were less purified were selected to be low in nickel. With chicks housed in an ultra-clean environment it was possible to produce low concentrations of fat in the liver, abnormal pigmentation of the skin, swelling of the hock joint, and thickening of the long bones, and in one experiment there was reduction in growth (45,46). Supplementation of the diet with 3 or 5 mg nickel per kilogram of diet prevented all of these abnormalities.

Evidence that vanadium may have an essential function was reported for young chicks maintained in an ultra-clean environment (47). The purified casein diet contained less than 10 ppb vanadium. Supplemented birds received drinking water containing 2 ppm vanadium from NH_4VO_3. The growth rate of the primary wing feathers and tail feathers were retarded and vanadium concentration of the liver, kidney, and heart were reduced in deficient birds compared to those receiving vanadium. At 28 days of age, the concentration of plasma cholesterol was lower in deficient birds than in vanadium-supplemented birds. By 49 days of age this relationship was reversed.

By the use of highly controlled environmental systems it is likely that requirements for additional elements will be demonstrated. Many of the elements in the periodic table not presently known to be essential can be considered as candidates. Bowen has stated that only actinium, plutonium, polonium, protactinium, and radium should presently be considered non-essential, on the basis that they occur in organisms at concentrations below 1 atom per cell (48). Liebscher and Smith have proposed a means of distinguishing between essential and non-essential elements (49). Their method is based on evaluation of the distribution pattern of elemental concentrations in tissues of population groups. Differentiation is based on existence of homeostatic control mechanisms for essential elements only. Pitfalls in this approach were discussed.

It is particularly important that investigators not be blinded by the presently recognized and emphasized toxicity of many metals. It is likely that at lower levels some of the current "toxic" metals may indeed be required for life itself. For fluoride and selenium the quantities separating desirable and undesirable levels of intake are much less than one order of magnitude. In addition to demonstrating gross beneficial effects of elements under given experimental conditions, it is mandatory to establish for each element the precise mechanism of the physiological effects and functional relationship to other nutrients, stage of the life cycle and medical and environmental stresses. Apart from reducing toxic exposures, this type of information is required for modifying human intake at lower levels.

SUMMARY AND CONCLUSIONS

Several metals and other inorganic elements are required by human beings in amounts ranging from a few micrograms to 18 milligrams per day for iron. For some essential trace elements, we know a great deal about the requirement and the effects of deficiency in man. For others, our understanding is fragmentary and based largely on analogies to effects in animals. Evidence for requirements of tin, nickel, and vanadium has recently been obtained in experimental animals and other "new" required elements are anticipated.

Toxic levels of both essential and non-essential elements are known to exist in man's environment today. It is known that some of the toxic levels interfere with the essential functions. In this context the nutritional state of the population is important for resisting the adverse effects of environmental contaminants. In some instances, we know that the effects of a toxicant may be increased by low intake of one or more key nutrients or that the toxicity may be lessened by intake of nutrients in excess of the generally recognized requirement. Food additives such as synthetic chelating agents and amino acids may also alter the nutrient-toxicant relationship. These are areas that need further investigation.

REFERENCES

1. Six, K. M. and Goyer, R. A. (1970). Experimental enhancement of lead toxicity by low dietary calcium. J. Lab. Clin. Med. 76: 933.

2. Smith, J. C. and Schwarz, K. (1967). A controlled environment system for new trace element deficiencies. J. Nutr. 93: 182.

3. Underwood, E. J. (1971). Trace Elements in Human and Animal Nutrition. (3rd ed.) Academic Press, New York.

4. National Academy of Sciences. (1968). Recommended Dietary Allowances. (7th ed.) Publication No. 1694. Washington.

5. Dowdy, R. P. (1969). Copper metabolism. Am. J. Clin. Nutr. 22: 887.

6. Peisach, J., Aisen, P. and Blumberg, W. E. (eds.) (1966). The Biochemistry of Copper. Academic Press, New York.

7. Prasad, A. S. (ed.). (1966). Zinc Metabolism. Charles C. Thomas, Springfield, Illinois.

8. Schroeder, H. A. et al. (1967). Essential trace metals in man: Zinc. Relation to environmental cadmium. J. Chron. Dis. 20: 179.

9. Prasad, A. S. (ed.). (1969). Symposium on zinc metabolism. Am. J. Clin. Nutr. 22: 1215, 1279.

10. Fox, M. R. S. (1970). The status of zinc in human nutrition. World Rev. Nutr. and Dietet. 12: 208. Karger, Basel.

11. Schroeder, H. A., Balassa, J. J. and Tipton, I. H. (1966). Essential trace metals in man: Manganese. J. Chron. Dis. 19: 545.

12. Cotzias, G. C. (1967). Importance of trace substances in environmental health as exemplified by manganese. In: D. D. Hemphill (ed.) Trace Substances in Environmental Health, vol. 1, University of Missouri: Columbia.

13. Shaw, J. H. (1967). Present knowledge of fluoride. In: Present Knowledge in Nutrition, (3rd ed.). The Nutrition Foundation, Inc., New York, p. 130.

14. Adler, P. et al. (1970). Fluorides and Human Health. WHO Monog. Series No. 59, World Health Organization, Geneva, Switzerland.

15. Levander, O. A. (1967). Present knowledge of selenium. In: Present Knowledge in Nutrition, (3rd ed.). The Nutrition Foundation, Inc., New York.

16. Muth, O. H., Oldfield, J. E. and Weswig, P. H. (ed.). (1967). Selenium in Biomedicine. Avi Publishing Company, Westport.

17. Schroeder, H. A. (1968). The role of chromium in mammalian nutrition. Am. J. Clin. Nutr. 21: 230.

18. Mertz, W. (1969). Chromium occurrence and function in biological systems. Physiol. Rev. 49: 163.

19. Blanc, B. et al. (1968). Nutritional anemias. Tech. Rpt. Series No. 405. World Health Organization, Geneva, Switzerland.

20. Fox, M. R. S. and Fry, B. E. (1970). Cadmium toxicity decreased by dietary ascorbic acid supplements. Science 169: 989.

21. Murphy, E. W., Page, L. and Watt, B. K. (1971). Trace minerals in type A school lunches. J. Am. Dietet. Assoc. 58: 115.

22. Starcher, B. C. (1969). Studies on the mechanism of copper absorption in the chick. J. Nutr. 97: 321.

23. Carnes, W. H. et al. (1968). The role of copper in the circulatory system. In: D. D. Hemphill (ed.). Trace Substances in Environmental Health, vol. 2, Univ. of Missouri: Columbia.

24. McCall, J. T. et al. (1967). Comparative metabolism of copper and zinc in patients with Wilson's disease (hepatolenticular degeneration). Am. J. Med. Sci. 254: 13.

25. Henkin, R. I. (1969). Trace metals and taste. In: D. D. Hemphill (ed.). Trace Substances in Environmental Health, vol. 3, University of Missouri, Columbia.

26. Tucker, H. F. and Salmon, W. D. (1955). Parakeratosis in zinc deficiency disease in pigs. Proc. Soc. Exp. Biol. Med. 88: 613.

27. Prasad, A. S. (ed.) (1966). Metabolism of zinc and its deficiency in human subjects. In: Zinc Metabolism, Charles C. Thomas, Springfield, Illinois.

28. Ronaghy, H. A. et al. (1970). A preliminary report on zinc supplementation. Symposium on Food Science and Nutritional Diseases in the Middle East. (abstract).

29. Ronaghy, H. et al. (1969). Controlled zinc supplementation for malnourished school boys: A pilot experiment. Am. J. Clin. Nutr. 22: 1279.

30. Halsted, J. A. (1968). Geophagia in man: Its nature and its nutritional effects. Am. J. Clin. Nutr. 21: 1384.

31. Smith, J. C. Jr. and Halsted, J. A. (1970). Clay ingestion (geophagia) as a source of zinc for rats. J. Nutr. 100: 973.

32. Hurley, L. S. (1968). Approaches to the study of nutrition in mammalian development. Fed. Proc. 27: 193.

33. Henkin, R. I. (In press). Newer aspects of zinc and copper. In: W. Mertz (ed.). Newer Trace Elements in Nutrition, Marcel Decker, New York.

34. Pories, W. J. et al. (1967). Acceleration of healing with zinc sulfate. Ann. Surg. 165: 432.

35. Pories, W. J. et al. (1967). The treatment of atherosclerosis with zinc sulfate. Military Med. Sect., American Medical Association Meeting. (abstract).

36. Marjanen, H. (1969). Possible causal relationship between the easily soluble amount of manganese on arable mineral soil and susceptibility of cancer in Finland. Ann. Agric. Fenn. 8: 326.

37. Anon. (1968). Epidemic cardiac failure in beer drinkers. Nutr. Rev. 26: 173.

38. Dobrenz, A. R. et al. (1964). Effect of a minimal fluoride diet on rats. Proc. Soc. Exp. Biol. Med. 117: 689.

39. Thompson, J. N. and Scott, M. L. (1970). Impaired lipid and vitamin E absorption related to atrophy of the pancreas in selenium deficient chicks. J. Nutr. 100: 797.

40. Hopkins, L. L. Jr. and Majaj, A. S. (1967). Selenium in human nutrition. In: O. H. Muth, J. E. Oldfield and P. H. Weswig (ed.). Selenium in Biomedicine. Avi Publishing Company, Westport.

41. Halverson, A. W., Hendrick, C. M. and Olson, O. E. (1955). Observations on the protective effect of linseed oil meal and some extracts against chronic selenium poisoning in rats. J. Nutr. 56: 51.

42. Levander, O. A. and Morris, V. C. (1970). Interactions of methionine, vitamin E, and antioxidants in selenium toxicity in the rat. J. Nutr. 100: 1111.

43. Hambidge, K. M. (In press). Chromium nutrition in the mother and growing child. In: W. Mertz (ed.) Newer Trace Elements in Nutrition, Marcel Decker, New York.

44. Schwarz, K. D., Milne, D. B. and Vinyard, E. (1970). Growth effects of tin compounds in rats maintained in a trace-element controlled environment. Biochem. Biophys. Res. Com. 40: 22.

45. Nielsen, F. H. and Sauberlich, H. E. (1970). Evidence of a possible requirement for nickel by the chick. Proc. Soc. Exp. Biol. Med. 134: 845.

46. Nielsen, F. H. (In press). Studies on the essentiality of nickel. In: W. Mertz (ed.). Newer Trace Elements in Nutrition, Marcel Decker, New York.

47. Hopkins, L. L. Jr. (In press). The biological importance of vanadium. In: W.

Mertz (ed.). Newer Trace Elements in Nutrition, Marcel Decker, New York.

48. Bowen, H. J. M. (1966). Trace Elements in Biochemistry. Academic Press, New York.

49. Liebscher, K. and Smith, H. (1968). Essential and non-essential trace elements. Arch. Env. Health 17: 881.

CHAPTER 9. CASE FINDING:
USES AND LIMITATIONS

J. JULIAN CHISOLM, Baltimore City Hospitals
and Johns Hopkins University School of Medicine,
Baltimore, Maryland

Epidemiology, that is to say the collection and evaluation of statistical data on the distribution of disease and its attendant circumstances in a defined population, is most effective when there is presumptive evidence of exposure to the causative agent, when the effects of the agent are well established, and when the incidence of both the causative agent and of the effects can be measured with some precision. It presupposes a fairly sophisticated degree of knowledge about the agent, its effects, and methods of measurement.

When new or previously unidentified agents are under consideration, when their effects are indefinite or indistinguishable from those of other agents, or when adequate measuring techniques are not available, the epidemiological approach is badly handicapped. Some other approach to the detection of cause and effect relationships and the establishment of a hazard's magnitude is required.

Case-finding is the name given to one such procedure that has repeatedly demonstrated its usefulness in the elucidation of metallic poisoning problems.

Historically, recognition of the occurrence of human disease due to metal contaminants has followed a similar sequential pattern: Initially, an observant physician -- often an intern or medical student -- recognizes and reports a severe case of illness or a clustering of cases with similar clinical features. Although the clinical features are similar within the group, they present some subtle difference from the pattern of illness prevalent in the community. The subtle difference is what alerts the physician. He is thus stimulated to investigate further and this subsequent investigation may point to a heavy metal. This, in turn, may stimulate appropriate experimental studies in animals, the development of adequate laboratory techniques and further case-finding in the community.

Although case-finding alone cannot bring about effective control of a particular hazard, a thorough study of a clustering of cases of illness from the medical, biological, behavioral, and ecological viewpoints can help to delineate the extent and significance of a particular hazard and can help to provide some initial guidelines for effective means of control. In some instances, such study may indicate that control at the source may be quick, practical, and effective. In other instances, the metallic contaminant may be so widely disseminated in the environment that systematic screening of high-risk groups in the population will be necessary until appropriate control measures can be brought to bear on the problem. Finally, past experiences with the varied exposures to lead and other metals may, in turn, provide a basis for the development of more effective case-finding techniques in the future.

LEAD

The importance of the case-finding procedure is well demonstrated by studies of non-industrial lead poisoning.

The symptoms of mild lead intoxication (poor appetite, vomiting, constipation, irritability) are the same as those of many minor illnesses, while the clinical signs and symptoms of acute lead encephalopathy can resemble those caused by brain abscess, brain tumor, and some of the viral and bacterial infections of the brain and meninges. Except in classical cases of heavy metal poisoning which are either severe or very chronic, physical examination and routine laboratory examination of blood and urine ordinarily do not reveal information which points directly to a specific metal as the causative agent. Special laboratory tests are needed

In order to document both the level of exposure and absorption of the metal (level of the metal in blood, excreta and environmental source) and some adverse metabolic or functional impairment attributable to an excess of the metal. Often these essential laboratory techniques are not readily available in the general medical community. In severe cases, treatment may not fully reverse the injury to the tissue, particularly injury to the nervous system.

Lead Paint Poisoning in Children

BURNING OF BATTERY CASES -- The "Baltimore experience" may be taken as a model of the "case-finding" approach to this particular problem. Since 1914, when Thomas and Blackfan (1), working at The Johns Hopkins Hospital, were the first to point out in the American pediatric literature the occurrence of lead encephalopathy in children, the staff of that institution has been alert to the possibility of childhood plumbism. In 1932, for example, a child of seven was brought unconscious to The Johns Hopkins Hospital with a history of sudden onset of convulsions and stupor. The child had both a positive tuberculin test and a chest X-ray with findings characteristic of mediastinal tuberculosis, but lumbar puncture revealed that the spinal fluid was more typical of lead poisoning than of tuberculosis. With further diagnostic study, an astute intern correctly established the diagnosis of acute lead encephalopathy as the cause of the child's stupor and convulsions. Furthermore, she visited the home to determine the source of lead and alerted the Baltimore City Health Department. At the home, this intern learned that the father was dead and she found that the mother was also suffering from lead encephalopathy. A man in the house, Melrose Easter, pointed to the old wooden battery casings being used as fuel as a possible cause of the sickness. He further told them that other poor people in the neighborhood were using old wooden storage battery casings for fuel in their homes. Later, the physicians learned that this was a common practice among the poverty-stricken residents of the city during the depression days of the 1930's. Upon learning that the discarded wooden battery casings were being distributed either free or at very little cost to the poor by certain junk dealers, the Health

TABLE 1 -- LEAD PAINT POISONING IN CHILDREN
(Baltimore, Maryland 1931-1969[a])

Years	Numbers of Children			Deaths	
	A Asymptomatic Increased Lead Absorption[b]	B Lead Poisoning	C Total of A and B	Number	% of Cases of Poisoning
1931-44	-	145	145	63	43%
1945-49	-	97	97	17	18%
1950-54 [c]	-	220	220	25	12%
1955-59 [c]	399	338	737	19	5.6%
1960-64 [c]	350	232	582	10	4.3%
1965-69	215	111	326	2	1.8%
Totals		1,143		136	

a) Compiled from records of Baltimore City Health Department. (Age distribution: 12-48 months of age - 94% of all cases of poisoning)
b) Asymptomatic Increased Lead Absorption = blood lead concentration ≥ 60 µg Pb/100 gm whole blood. Data not available prior to 1954.
c) Individual pilot screening programs during each of these periods.

Department effectively stopped the distribution, thereby terminating the outbreak. A total of 40 such cases of plumbism in children and adults were diagnosed in 1932 and reported in the medical literature (2). A similar outbreak occurred in Rotherham, England, due, in this instance, to the burning of discarded battery casings made of vulcanite (3,4).

Cases of plumbism in Baltimore due to the burning of storage battery casings quickly disappeared with the prompt enforcement of adequate control measures, but medical awareness of the disease picture had been stimulated.

PICA -- A free diagnostic service established by the Baltimore City Health Department in 1935 was used at an ever-increasing rate, with the result that there was a progressive increase in the numbers of cases recognized in children. As a result, it became apparent that the eating of leaded paint by children with pica was the principal cause of plumbism in young children in Baltimore. Between 1931 and 1940, 24.3% of all cases of childhood plumbism for the entire United States were reported from Baltimore (5). This apparent unusual concentration of such cases in a single city was most probably a reflection of a high level of medical awareness of the entity in the city and the availability of the only free public health diagnostic service in the country at the time. Table 1, showing the numbers of reported cases and deaths in Baltimore between 1931 and 1969, indicates that, at first, only severe cases were recognized; indeed, between 1931 and 1944, 63 or 44% of the reported cases died. Subsequently, with improved recognition and improved methods of treatment, the numbers of cases increased while the apparent mortality decreased. The disease was made reportable. Beginning in 1949, the Baltimore City Health Department assigned a single visiting Public Health Nurse to the full-time duty of investigating all cases in children reported to the Health Department. Ever since that time, the "lead nurse" has visited each home to obtain a medical history, to insure examination of all other young children in the home and to obtain paint samples for analysis in the Health Department laboratory.

Baltimore has a convalescent children's facility where children may remain in safety until their homes are inspected to ensure complete removal of the paint or until a new safe dwelling is found for the family. Furthermore, all housemates with similar exposure are tested and usually one or more additional children with either lead poisoning or increased lead absorption are so detected.

Although most cases of childhood plumbism are today found in dwellings built prior to 1940 and such dwellings today contain the largest portion of buildings classified as "deteriorating" or "dilapidated," lead pigment paints for interior use continued to be sold in Baltimore until 1958 when an appropriate local ordinance was enacted prohibiting further sale of such paints for interior use in the city. The total experience in Baltimore suggests that plumbism in children with pica is likely to continue for some years to come unless the present approach can be modified.

These observations point up the limitations of the "case-finding" approach. A continuation of "classical" cases may be anticipated so long as the hazardous buildings are occupied. Deaths will be few, but far more significantly, each year new instances of permanent brain damage due to childhood plumbism will continue to occur. Without intensive screening, of course, the true incidence of the disease in the community and its sequelae will not be known. In Baltimore, any further improvement in the

protection of child health will require a
reorientation of attitude and public policy.
Large-scale screening of both children and their
dwellings, coupled with better enforcement of
existing regulations, will be necessary until the
housing hazard is finally eliminated.

The "Chicago experience" provides another
interesting example. In 1953, a U.S. P.H.S.
medical officer assigned to the Communicable Disease
Center, was dispatched to Chicago to investigate a
serious outbreak of poliomyelitis. This physician
was sensitized to childhood plumbism, having just
completed his internship in pediatrics at The Johns
Hopkins Hospital in Baltimore. He recognized that
several of the "polio" cases were not typical of
poliomyelitis; rather, they seemed to him to fit
better the clinical picture of plumbism.
Investigation followed and in 1953, 13 cases of lead
intoxication were documented at a single hospital.
At this hospital, only 20 such cases had been
recognized during the preceding 14 years. By 1954,
the disease was made reportable and the laboratory of
the Chicago Board of Health began providing the
necessary laboratory diagnostic service. Table 2
summarizes data for the 1960's, in Chicago. Alarmed
by an unacceptable death rate due to childhood
plumbism in the early 1960's, Chicago, with Federal
assistance, began a mass screening of children (6).
Five and seven tenths per cent of those tested (Table
3) were found to have increased lead absorption
(blood lead concentration 50 µg Pb/100 gm whole
blood) and 1,154 children or 1.65% of those tested

TABLE 2 -- LEAD PAINT POISONING IN CHILDREN
(Chicago, Illinois 1960-69) (7)

Years	Number of Cases of Poisoning		Deaths	Deaths/year
	Total	Cases/Year		
1960-65	1081		110	
		180		18.3 (10%)
1966	320		7	
		320		7 (2%)
1967-68	1336		17	
		668		8.5 (3%)
1969	450		1	
		450		1 (0.2%)

TABLE 3 -- DATA OF SURVEY AREAS FOR CHICAGO BLOOD LEAD SCREENING PROGRAM -- 1967-1968 (6)

Urban Area	Number of Children Under 5 years (1960 Census) (A)	1967				1968				Totals for 1967-1968			
		Number Tested (B) a	% of A	Number with high lead b	% of B	Number Tested (C) a	% of A	Number with high lead b	% of C	Number Tested (D) a	% of A	Number with high lead b	% of D
1	10,000	2,500	25	157	6.3	3,100	31	53	1.7	5,600	56	210	3.8
2	15,100	1,700	11	73	4.3	1,800	12	56	3.1	3,500	23	129	3.7
3	22,000	2,000	9	262	13.1	3,000	14	166	5.5	5,000	23	428	8.6
4	22,300	5,200	23	509	9.8	5,500	25	270	4.9	10,700	48	779	7.3
5	13,700	4,100	30	316	7.7	4,900	36	142	2.9	9,000	66	458	5.1
6	24,500	6,600	27	393	6.0	7,700	31	277	3.6	14,300	58	670	4.7
7	20,500	3,400	17	432	12.7	7,000	34	356	5.1	10,400	51	788	7.6
8	15,300	2,200	14	218	9.9	6,100	40	227	3.7	8,300	54	445	5.4
9	4,500	300	7	19	6.3	1,700	38	9	0.5	2,000	44	28	1.4
Totals	147,900a	28,000a	19	2,379	8.5	40,800a	28	1,556	3.8	68,800a	47	3,935	5.7

a) Figures given to the nearest hundred.
b) High lead = 50 µg lead/100 ml whole blood or higher.

215

were treated as cases of lead intoxication. Deaths dropped to the vanishing point and a treatment center was set up to follow the affected children on an out-patient basis (7), this being necessary because of difficulties in obtaining effective elimination of the housing hazard. The outcome in the children treated in the presence of continued pica and environmental exposure will not be known for several years. However, the data of Byers and Lord (8) and Perlstein and Attala (9) indicate that recurrent plumbism, even in the absence of known encephalopathy, may ultimately lead to permanent brain damage in 10% or more of the affected children. The "child-screening" approach alone does not exert adequate etiologic control. The experience of these two cities (Baltimore and Chicago) indicate that large-scale screening of both houses and children, coupled with prompt enforcement of regulations requiring the removal of leaded paints in the houses of the affected children, will be necessary until this type of hazard in old housing is finally eliminated. The use of leaded paints in houses now being built must be controlled in such a manner that hazards to children from this source are prevented in the future.

Plumbism from Lead-glazed Earthenware

Cases of lead poisoning have for centuries been traced in all parts of the world to the use of improperly lead-glazed earthenware food and beverage containers. Acidic foods and beverages such as alcoholic beverages, cola drinks (10), fruits, fruit juices, and tomatoes contain acetate, citrate, and malate which form soluble salts with lead. If the lead silicate glaze on the vessel has not been properly fired and fused during fabrication, these acidic foods can leach the lead from the glaze into the food, which is then consumed. A recent report of plumbism from this source is that of Klein et al. (11), in Montreal, Canada. The report details the cases of two children, one of whom died, who drank apple juice from an earthenware pitcher with a lead glaze. More important, these investigators went a step further, purchased and tested 117 earthenware vessels intended for use as food and beverage containers. They also prepared and tested in the McGill University ceramics laboratory 147 samples, using 49 different glaze compositions commonly used by various schools, clubs, studios, and potters. The results are shown in Table 4. The United States Potter's Association and the United States Food and Drug Administration have defined 7 ppm as the maximal lead release for glazes recommended for use on ceramic items intended for food and drink. On this basis, 132 or 50% of the 264 samples tested were unsafe by the 7 ppm standard and 10 to 25% were grossly unsafe. The data in Table 4 indicate that commercial ware might be just as unsafe as home handicrafted and imported ware. Whether the 7 ppm limit provides an adequate safeguard should be subject to reevaluation, as it does not necessarily take into account the acidity (cola drink=pH 2.7) of the beverage, temperature and quantity of beverage consumed, nor the length of time an acidic food or beverage is in contact with the glaze. All of these factors can influence the quantity of lead finally ingested by the user. It would be impractical to conduct epidemiological investigations of a nation's population to pick out the small proportion who might have used contaminated pottery. One has the choice, either of identifying the source of lead in afflicted people, i.e., from case-finding procedures, or of preventing the distribution of improperly prepared pottery (which has not been entirely successful to date).

Plumbism from Illicit Whiskey

A somewhat comparable situation exists here. History records the fact that as early as 1620, the Arabians used distilled alcoholic beverages. The Spaniards learned the art from them and passed it on to the Irish monks. Ireland was, at the time, under British rule and the British imposed an excise tax on distilled beverages. And so the first attempts to evade the tax occurred in the British Isles. Thus was born the practice of making illicit or "moonshine" whiskey. Indeed, in 1794, in the United States there was an open insurrection known as the "Whiskey Rebellion," over payment of a tax on alcoholic beverages. Wherever Scottish or Irish immigrants settled in the United States, moonshine whiskey making has flourished. Today, this clandestine "industry" is concentrated in the southeastern part of the United States. According to one U.S. Treasury Agent, over 90% of the samples of

TABLE 4 -- REPORT OF LEAD EXTRACTIONS FROM 264 EARTHENWARE GLAZE SURFACES (11)

Sample Group	No. of Glaze Surfaces	Lead Extraction (ppm)						
		<0.5	0.5-7	7-20	20-100	>100	100-500	500-1000
Group A								
Imported pottery	29	9	5	6	6	3	-	-
Domestic commercial ware	48	17	7	4	15	5	-	-
Domestic handcraft	40	10	4	4	6	16	-	-
Totals	117	36	16	14	27	24		
Group B	147	36	36	32	14	-	10	19

Group A - based on analysis of samples obtained over-the-counter from handicraft shops and department stores

Group B - samples of identical size and clay composition treated with 49 different glaze compositions

N.B. Standard room temperature acetic acid test used - Interpretation <7 ppm Pb considered safe for table use

218

confiscated moonshine whiskey contain "high concentrations of lead" (12). In the illicit manufacture of moonshine, discarded automobile radiators which contain lead may be used as condensers and lead solder is used in the seams and joints of the tubing of the distillation unit. From these sources, it is leached into the final product probably largely as lead acetate. Medical reports attest to the presence of severe lead poisoning in this area among the consumers. Thus, lead encephalopathy (13), chronic lead nephropathy with secondary gout (14), and more subtle impairment of hormonal and renal function (15) have been reported. (The chronic lead nephropathy seen in these long-term illicit whiskey drinkers is similar to that first reported by Nye (16) in Australia among survivors of chronic childhood plumbism. Its clinical features and diagnosis have been well described by Emmerson (17).) Physicians caring for the patients in the southeastern United States report that the population affected is not limited to the chronic skid-row alcoholic; rather, consumers include weekend drinkers, social drinkers, and well-to-do adults for whom cost is no object but who state simply that they prefer "moonshine" to the legal (safely-distilled) product. Conventional attempts to control this through law enforcement and through warnings to the public in the news media have not been eminently successful. It is estimated that production in the southeastern United States continues at a rate of millions of gallons annually. Presumably, it is consumed! Nevertheless, there is no systematic screening of the population at risk. Medical interest seems focused mainly on the toxic effects of the product. Furthermore, as with children, clinical diagnosis is difficult and uncertain except for those few physicians interested in and familiar with plumbism. Indeed, accurate clinical diagnosis is further confounded by the fact that the symptoms of acute alcoholism and acute lead intoxication share many common features. Screening, therefore, seems indicated perhaps on the basis of adult hospital admissions, emergency room visits, prenatal examinations or annual medical examination, in order to obtain an estimate of the extent of this particular hazard to human health.

Plumbism from Unusual Sources

Sporadic cases of plumbism in children and adults have been traced to the following: lead-painted children's toys and furniture, lead toys and baubles eaten by children, lead nipple shields, home battery manufacture, artist's paint pigments (hand-mixing), lead dust in shooting galleries (attendant at risk), soft well water conveyed in lead pipes, ashes and fumes of painted wood and discarded battery casings used for fuel in stoves and fireplaces, "medications," jeweler's waste, and lead type in schools for the blind. Lead pigments have also cropped up through the years in cosmetics. Most of these sources have or can be controlled at the source, so that today, they are not major sources to which large segments of the human population are exposed. Nevertheless, continued surveillance, control at the source and education of the limited population at risk are necessary. Lead poisoning in dogs, monkeys (18), cattle, and ducks are also well-known entities. Aside from the economic and esthetic impact of plumbism in animals, investigation and reporting of new outbreaks of disease suggestive of metal poisoning in animals can serve as an early warning system of similar hazards to humans. Human ingenuity continuously finds new uses for metals, including lead; and the hazards of certain usages of lead recognized in past years, but since forgotten, can result in recurrence of such human hazards today. Case-finding and screening for plumbism in animals and humans serve an important early warning system for the protection of human health.

BERYLLIUM

The usefulness and limitations of the case finding technique are well illustrated by the history of the Beryllium Case Registry, now being operated by the Pulmonary Unit of the Massachusetts General Hospital under contract to the National Institute of Environmental Health Sciences. (See also Chapter 5)

Extraction of the ores and processing the metal in the United States started in 1936. Routine medical surveillance early identified acute affections in the form of dermatosis from contact with soluble salts of beryllium, and acute respiratory responses to fumes from sublimation processes. "In the latter part of 1944, a chronic type of pulmonary disease, presenting a roentgenologic pattern heretofore not described, expressed itself in a resident in the immediate vicinity of the beryllium producing plant...... We have a total of 13 cases under study presenting pneumonic pathology classified as granulomatosis. Of this group, four are ex-employees of the plant..... A mass chest X-ray survey in June 1948,......covered over 10,000 people and elicited two suspected non-employee cases..... No in-plant cases were discovered despite the fact that the majority were employees of long standing and exposed to relatively heavy concentrations of beryllium compounds in comparison to the potential neighborhood exposure as shown in air pollution studies"(19).

This extract from the Introduction to a conference held in 1966 indicates the uncertainty that attended these early cases, and the failure of a preliminary epidemiologic study to clarify the issue. The situation proved to be even further complicated by the fact that the roentgenologic picture of the chronic case is almost indistinguishable from that of sarcoidosis, a disease occurring apart from exposure to beryllium, whose real etiology is still obscure (20).

The evidence was sufficient, however, to bring about the institution of industrial controls in 1949. In 1951 a Beryllium Case Registry was started by Dr. Harriet Hardy (21) to secure data which would answer eight questions:

1. What are the criteria for the diagnosis of beryllium poisoning in both its acute and chronic form?

2. What was the quality of work exposure in cases of beryllium poisoning?

3. What was the quantity of beryllium compounds causing disease?

4. How effective have the controls introduced in 1949 proved to be?

5. What is the clinical course of beryllium disease, its prognosis, and its complications?

6. What has been the response to steroid therapy introduced in 1949?

7. How many workers and neighbors were exposed to similar risks and did not get ill?

8. As in animal studies, do workers with and without diagnosable beryllium disease suffer a significant increase in malignancy, especially of the bone or lung?

The personal advocacy of Dr. Hardy, supplemented by appeals in relevant medical journals, has resulted in the collection of over 700 cases. A complete case record would include occupational history, medical history, radiographs, tissue analysis for beryllium, histologic slides, and urinalysis for beryllium. Pulmonary function studies have been carried out on a limited number (22).

From the files an excellent picture can be developed of the clinical features of chronic pulmonary berylliosis and its complications, and earlier ideas on the progress of the disease have been corrected. The continued development of new cases, some as a result of recent exposure and some after a long delay from old exposure, has been demonstrated. The existence of dangers from newer uses of beryllium compounds and alloys is evident, and neighborhood cases continue to be revealed. The possibility of late carcinogenic effects is under study, but the numbers are yet too small for definite conclusions (23).

Unfortunately, however, the records do not lend themselves to extensive epidemiologic analysis. Few of the records include all of the items listed above. Because reporting is entirely voluntary, no reliable estimate is possible of the total number of cases that have developed. The extent of the population at risk has also become progressively more uncertain as the use of beryllium compounds and alloys has diversified, and involved small business enterprises some echelons removed from the big producers who warn their first line purchasers of the risks. In few cases is the actual extent of the individual's exposure known, although it is evident from the decline in the number of cases with first exposure occurring after 1949 that the industrial controls instituted at that time must have been quite effective (24,25).

One indirect benefit to be obtained from maintaining a registry of this character is the existence of a specialist staff to which a practitioner can turn for diagnostic assistance or from which a troubled community group can get guidance on seeking investigation. Both aspects have been served by the Beryllium Case Registry.

CONCLUSIONS

In virtually all instances of metallic poisoning among non-industrial groups, the initial suspicion that the disease state might result from some metallic contaminant can be traced to an astute clinical observation. This somewhat fortuitous approach depends very heavily on the alertness, acumen and diligence of individual physicians; but it is essential for the initial recognition of unsuspected environmental risks. To be effective, it must be followed up by comprehensive epidemiologic evaluation. By itself, case finding exerts only weak preventive influence to control the identified hazard.

Examples are cited throughout this book in which thorough study of a clustering of cases of illness has indicated different approaches to the control of a particular hazard. The epidemiology of Minamata disease (Chapter 2) pointed to prevention of further disease through surveillance of fish and other potential aquatic vectors, while further contamination of the environment by methyl mercury is eliminated by control at potential sources. Lead poisoning due to the burning of battery casings (Chapter 3) can be eliminated through control of their distribution and by their disposal in a safe manner. On the other hand, the wide-spread distribution of leaded paints in residential interiors, the long life and slow rate of deterioration of old housing, indicate that plumbism in children with pica cannot be quickly eradicated through controls exerted at source and distribution points. Prevention of this particular form of lead exposure apparently calls for systematic screening of children and housing in high-risk areas until the hazard in old housing is finally eliminated and the use of safe paints in current housing is assured.

Medical case reports of suspected or proven metal poisonings can provide an important early warning device. They serve to alert the community to recurrences of hazardous exposures long forgotten, as well as to potential new but unsuspected hazards related either to new uses of metals or new distribution patterns among the general population. Other possible alerting devices might be more effectively utilized. For example, bacteriologic studies generally are included as a standard part of an adequate post mortem examination, even though infectious agents are not necessarily considered a significant component of the terminal illness. On the other hand, analysis of tissues for heavy metals is rarely considered at post mortem examination except on very strong suspicion of an acute toxic overdose. The frequency of chronic nervous system injury in a variety of metal poisonings suggest that analysis of tissues for metals may yield useful information in cases in which the picture of illness is compatible with heavy metal poisoning. Outbreaks of illness in wild and domesticated animals, if studied with respect to metals and evaluated in terms of their relevance to human exposure to the same metal, can serve as still another early warning device.

Past experience clearly shows that the identification and solution of environmental problems requires the cooperation of individuals with a broad span of interests and skills. Health professionals, through their knowledge of disease patterns, can alert the community through case-finding techniques. The interdependence of many disciplines and the need for communication and cooperation among these disciplines is self evident. No one of them acting alone can formulate or effect an adequate solution.

It will be evident that case-finding, often constituting a first approach to the elucidation of the true etiology of a new syndrome, seldom continues as the sole method. As the possibilities of causation are narrowed, and cause-effect relationships are glimpsed, more deliberate and precise procedures become possible. Epidemiology in particular can be undertaken with more confidence when the probabilities of success can be gauged. In some of the instances cited in this chapter it would be difficult to say just when the procedures passed from the probing procedures of case-finding to the directed collection of statistical evidence. This is as it should be; the name of the game is of much less importance than the game itself.

REFERENCES

1. Thomas, H. M. and Blackfan, K. D. (1914). Recurrent meningitis, due to lead, in a child of five years. Am. J. Dis. Child. 8: 377.

2. Williams, H. et al. (1933). Lead poisoning from the burning of battery casings. J. Amer. Med. Assoc. 100: 1485.

3. Gillett, J. A. (1955). An outbreak of lead poisoning in the Canklow District of Rotherham. Lancet 1: 1118.

4. Travers, E., Rendle-Short, J. and Harvey, C. C. (1956). The Rotherham lead-poisoning outbreak. Lancet 2: 113.

5. Williams, H. et al. (1952). Lead poisoning in young children. Public Health Reports 67: 230.

6. Blanksma, L. A. et al. (1969). Incidence of high blood lead levels in Chicago children. Pediatrics 44: 661.

7. Sachs, H. K. et al. (1970). Ambulatory treatment of lead poisoning: Report of 1,155 cases. Pediatrics 46: 389.

8. Byers, R. K. and Lord, E. E. (1943). Late effects of lead poisoning on mental development. Am. J. Dis. Child. 66: 471.

9. Perlstein, M. A. and Attala, R. (1966). Neurologic sequelae of plumbism in children. Clin. Pediat. 5: 292.

10. Harris, R. W. and Elsea, W. R. (1967). Ceramic glaze as a source of lead poisoning. J. Amer. Med. Assoc. 202: 544.

11. Klein, M. (1970). Earthenware containers as a source of fatal lead poisoning. Case study and public-health considerations. New Eng. J. Med. 283: 669.

12. Hughes, E. D. (1967). What doctors should know about moonshine. Resident Physician: 78.

13. Crutcher, J. C. (1963). Clinical manifestations and therapy of acute lead intoxication due to the ingestion of illicitly distilled alcohol. Ann. Intern. Med. 59: 707.

14. Morgan, J. M., Hartley, M. W. and Miller, R. E. (1966). Nephropathy in chronic lead poisoning. Arch. Intern. Med. 118: 17.

15. Sandstead, H. H. , Michelaskis, A. M. and Temple, T. E. (1970). Lead intoxication, its effect on the renin-aldosterone response to sodium deprivation. Arch. Env. Health 20: 356.

16. Nye, L. J. J. (1929). An investigation of the extraordinary incidence of chronic nephritis in young people in Queensland. Med. J. of Australia 2: 145.

17. Emmerson, B. T. (1968). The clinical differentiation of lead gout from primary gout. Arthritis and Rheum. 11: 623.

18. Sauer, R. M., Zook, B. C. and Garner, F. M. (1970). Demyelinating encephalomyelopathy associated with lead poisoning in non-human primates. Science 169: 1091.

19. DeNardi, J. M. (1966). Personal communication.

20. Hardy, H. L. (1956). Differential diagnosis between beryllium poisoning and sarcoidosis. Am. Rev. Tuberculosis Pulm. Dis. 74: 885.

21. Hardy, H. L., Rabe, E. W. and Lorch, S. (1967). United States Beryllium Case Registry. J. Occup. Med. 9: 271.

22. Andrews, J. L., Kazemi, H. and Hardy, H. L. (1969). Patterns of lung dysfunction in chronic beryllium disease. Am. Rev. Resp. Dis. 100: 791.

23. Mancuso, T. and El-Attar, A. A. (1969). Epidemiological study of the beryllium industry. Cohort methodology and mortality studies. J. Occup. Med. 11: 422.

24. Peyton, M. F. and Worcester, J. (1959). Exposure data and epidemiology of the Beryllium Case Registry. Arch. Ind. Health 19: 94.

25. Williams, C. R. (1959). Evaluation of exposure data in the Beryllium Registry. Arch. Ind. Health 19: 263.

CHAPTER 10. ANALYTICAL CONSIDERATIONS

LLOYD B. TEPPER, Kettering Laboratory
University of Cincinnati College of Medicine
Cincinnati, Ohio

The appraisal of the concentrations of the metals under consideration in man and in the environment is dependent upon accurate sampling and analysis of materials. This discussion presumes that the substances sampled are truly representative of areas or populations being investigated. While we shall not dwell on this point, it cannot be over-emphasized that samples must truly reflect what they purport to measure, and that the pattern of sampling cannot be the result of casual collections of whatever may seem convenient to collect. Some reports appearing in the scientific literature, for example, arrive at general conclusions about airborne metals from data collected at highly arbitrary and non-representative sites.

PRINCIPLES

Environmental metals must be collected and put into suitable form for analysis. Airborne metals are almost always particulate rather than gaseous. Exceptions are metallic mercury and certain organometallic compounds, such as simple organolead and organomercury compounds, which are volatile and may exist in atmospheres in a gaseous phase. Particulates can be collected on a filter medium through which air is drawn by a pump. In practice the use of a paper or paper-like medium is most common in ambient air sampling. Particulates can also be separated from the air by impingement in a liquid-filled collector or by precipitation of charged particles in electrostatic fields. Organolead or mercury compounds which exist as volatile non-particulate material may be extracted from the air by adsorption on charcoal or by direct reaction with a specific chemical reagent.

Metals in water or in solution following treatment with acid can be concentrated by evaporation. Metals in soils may require separation by chemical techniques prior to analysis.

In most traditional methods of analysis tissues or biological fluids require preliminary chemical treatment. Such treatment reduces the material to a clear liquid free of extraneous organic matter. The term "ashing" or digestion is applied to the procedure. Alkylmercury compounds and similar simple organometallics may be assayed directly with non-destructive methods if the substance can be volatilized from a collection medium or sample and directed into instruments such as the gas chromatograph.

In the course of sample collection, storage, and preparation, loss of metals or contamination by metals may occur. Volatile metals such as lead, mercury, or cadmium may be driven off by the excessive application of heat in the ashing procedure, so that wet ashing without the application of heat may be indicated. When control or replicate samples are being run, it is essential that they undergo the entire analytical process since errors are often associated with sample ashing and preparation. The simple introduction of control amounts of pure materials in the final step of the assay does not validate the entire process of analysis.

The problem of contamination of samples is of great importance when low-level assays are being conducted. Contaminants may be inherent in filter media or in chemical reagents used in analyses. The meticulous laboratory will verify the quality of these materials, aqueous diluents, and glassware washing. Special attention to general housekeeping is necessary to minimize problems of contaminant transfer around the analytical area.

Available assays for the elements in question require the utilization of an equipped laboratory and the availability of experienced personnel, so that they are rarely suitable for field use. The only exception applies to mercury vapor which is detectable at industrial levels by the mercury vapor detector. Portable equipment for the direct measurement of airborne metallic particulates is under consideration and development. Many proposed devices have the disadvantages of excessive weight, lack of stability and drift of calibration. Several wet chemical methods can be adapted to kit form for field use, especially when high industrial levels of contamination are anticipated. The procedures are most useful in ascertaining whether levels are above or below established operational criteria. From the practical point of view, the collection of particulates on filter medium for laboratory assay represents the currently recommended procedure.

It is generally true that the metallic content of liquid preparations derived from air, water, soil, or biological materials can be analyzed by several laboratory procedures. These can be categorized into two groups: wet chemical and instrumental methods. Wet chemical methods are based upon the development of a color, the intensity of which is measureable and related to the amount of the metal being assayed. This is the traditional approach, with the advantage of small capital cost for equipment but the disadvantages of limited capacity for numerous samples, the need for skilled technologists, and chemical interferences with the metal under examination, e.g.: other elements in the test solution which react in ways which lead to confusion or errors. Interfering metals can be removed by additional steps in sample preparation, but each added sample manipulation takes time and presents another opportunity for sample loss or contamination.

Instrumental methods of analysis reflect modern trends in this field and in electronics. In assays for metals the atomic absorption spectrophotometric technique (AAS) permits rapid, accurate analysis for most metals with relatively little sample preparation. The limiting factor in the productivity of a laboratory is not the instrument itself but rather the rate at which samples can be logged in, prepared, and reported. The capital investment required for AAS equipment ranges from $3,000-$12,000 depending on the degree of instrumental sophistication desired. Once the instrument is in operation, a careful technician with a minimum of special training can process the samples. A disadvantage of complex instruments is that highly specialized factory-trained technicians may be required to correct malfunctions which inevitably develop. In some regions it is not possible to obtain this kind of assistance, in which case the entire metal analysis capability of a laboratory may be destroyed for a prolonged or indefinite period. In such regions it is generally true that industrialization may be relatively under-developed. Consequently a small number of samples with limited variety may be more appropriately analyzed by older but perfectly satisfactory chemical methods.

Spectrographic techniques are suitable for certain metal analyses, and the choice of procedure often reflects the personal experience of those in charge of laboratory operations. The spectrographic technique is valuable in providing a tool which can search for and identify unknown metals in a sample. Colorimetric and atomic absorption methods described previously are used primarily when the specific element to be assayed is already known and a quantitative estimate is required.

Most analytical techniques are capable of determining the absolute amount of a metal present. A sample for analysis must be sufficiently large to provide at least the minimum quantity detectable by a particular procedure. This means that large samples are necessary when concentrations are low; high concentrations permit use of smaller samples. The assay itself does not indicate the metal concentration in the original air, water, or biological sample but rather only the quantity in the sample presented for analysis. If the weight or volume of the original sample is known, the concentration can then be expressed in terms of metal weight per unit volume (m^3, ml) or unit weight (g) of sample. For satisfactory results, samples should be sufficiently large to permit analyses of quantities of metal clearly in excess of the detectable minimal amounts.

Since the magnitude of a metal contamination problem is usually dependent upon concentration, it is clear that accurate chemical analysis must be preceded by accurate measurements of air or liquid volume or of sample weight. Problems are rarely associated with liquid or weight measurement; air volumes, however, are more difficult to measure. Instruments which measure air volume or air flow (volume per unit time, e.g.:m^3/min) must be calibrated against suitable standards. Under some measurement conditions corrections must be made to compensate for air volume changes due to unusual ambient temperatures or pressures.

Quality control invariably requires that blanks and standards be run with appropriate groups of test specimens. A blank is an assay sample composed only of a distilled water or an equivalent material which lacks the metal under investigation. As this blank sample is processed, it should yield an assay value of zero, or the scale should be adjusted so as to fix the zero at the point corresponding to the blank reading. High-reading blanks suggest contamination of reagents, collection media, or apparatus. Standards contain known quantities of the metal under investigation and are run in conjunction with blanks and the samples being assayed. The standards provide reference points at various concentration increments against which the test samples are compared.

Various kinds of errors are inherent in all analytical procedures, and replicate analyses of aliquots from a single sample do not yield precisely the same result. The inter-aliquot variability increases when the levels of the test metal are near the limit of detectability. High concentrations tend to give more uniform results. Laboratory error may also occur when operators make minor personal manipulative deviations from standard methods or when new sources of chemicals or other materials are used. Careful laboratory management tends to reduce errors and increase uniformity of performance; however, there is some element of error in all procedures. The presence of errors does not of itself indicate laboratory deficiencies. The meticulous laboratory will, however, continuously study the source and magnitude of errors and will be able to provide evidence of quality control and statistical estimates of variation in analysis.

233

It is appropriate to emphasize that the accurate assay of metals under discussion requires the highest level of experience and care. Hospital laboratories, small clinical laboratories, and certain kinds of mail-order analytical services may not be sufficiently familiar with metal assay to provide valid metal determinations. The fact that a laboratory has been certified to handle general clinical materials is not a sufficient guarantee of accuracy in metal studies. Experienced laboratories find it necessary to run a continuous program of internal validation of assays. Inter-laboratory comparisons on a national and international scale have been useful in identifying problems and in improving the quality of laboratory performance.

PREFERRED TECHNIQUES AND STANDARD VALUES

Analytical techniques commonly used for specific metals are summarized in the following discussion. Each method presented is suitable for the purpose indicated, and the degree of suitability under given operational circumstances will depend upon the experience of professional and technical personnel, the availability of equipment, and the volume, variety, and character of samples. Many methods have highly vocal advocates; in general each method is especially good in the hands of persons familiar with its applications and limitations.

In the reporting of urinary levels of metals, there has been reference to both Wt/1i and Wt/24 hours excretion. When the volume excreted in 24 hours equals 1.0 liter, the values in mg/1i and mg/24 hours are obviously the same. Usually, however, the volume excreted does not equal one 1i/24 hours. There is obvious merit to a 24-hour sample, but an accurate and valid collection over this period is difficult to obtain from persons with little interest in the assay or little experience in metabolic collections, i.e., almost everyone from whom a sample is likely to be obtained. The most common practice is to obtain a "spot" urinary sample and make the appropriate adjustment to permit the presentation of the data on a "per liter" basis. While a full 24-hour collection is desirable, regardless of whether the result is reported on a "per liter" or a "per 24 hour" basis, the "spot" sample may have validity in industrial screening situations. In some research applications, the fact that a full 24-hour collection has been obtained may be verified by a determination of urinary creatinine in the sample for metals. The daily excretion of creatinine is believed to be relatively constant and unrelated to the volume of urine excreted.

ARSENIC -- Most analyses are performed by the colorimetric method.

Approximate limit of detectability	1.0	μg
Air standard -- industrial (TLV)	0.5	mg/m^3
Water standard	0.05	ppm
Blood level -- normal	10.0	μg/100g
Urinary level -- normal	0 - 100.0	μg/li*
" " -- excessive	1000.0	μg/li
Hair -- normal	1.0	ppm

* May be higher if high amount of seafood in diet.

BERYLLIUM -- Standard methods utilize the spectrographic and colorimetric (morin) techniques. There has been some interest in the gas chromatographic procedure whereby beryllium is introduced into an organic molecule (chelate) which is sufficiently volatile for separation and measurement. An atomic absorption technique is also available; however, this is best used to distinguish between acceptable or non-acceptable industrial situations. It is not highly sensitive in beryllium assays where low-level or precise work is required.

Approximate limit of detectability	0.003	μg
Air standard -- industrial	2.0	μg/m^3
Air standard -- community	0.01	μg/m^3
Water standard	--	
Blood level -- normal	nil	
Urinary level -- normal	nil	
Tissue level, lung--normal	<0.1	μg/100g

CADMIUM -- Atomic absorption spectrophotometric and colorimetric procedures are satisfactory.

```
Approximate limit of
  detectability                                    0.01  µg
Air standard --  industrial
                 (TLV)                             0.2   mg/m³
  "        "    --  industrial
                 (TLV) for
                 cadmium
                 oxide fume,
                 not to
                 exceed                            0.1   mg/m³
Water standard                                     0.01  ppm
Blood level --  normal           0.3 -  5.0        µg/100g
  "      "  --  average                  0.85      µg/100g
Urinary level --  normal         0.5 - 11.0        µg/1l
  "         "  --  average               1.6       µg/1l
  "         "  --  excessive   Not established;
                               poor correlation
                               between level and
                               clinical status
```

CHROMIUM — Atomic absorption spectrophotometric and colorimetric procedures are satisfactory.

```
Approximate limit of
  detectability                                    0.1   µg
Air standard --  industrial
                 (TLV)                             0.5   mg/m³
  "        "    --  industrial
                 (TLV)
                 chromic
                 acid and
                 chromates                         0.1   mg/m³
  "        "    --  industrial
                 (TLV)
                 chromium,
                 soluble
                 chromic
                 and
                 chromous
                 salts                             0.5   mg/m³
Water standard                                     0.05  ppm
Blood level --  normal           1.0 -  5.5        µg/100g
  "      "  --  average                  2.8       µg/100g
Urinary level --  normal         1.5 - 11.0        µg/1l
  "         "  --  average               4.0       µg/1l
```

FLUORIDE* — Suitable fluoride preparations derived from air, water, or biological materials can be assayed by use of the fluoride electrode, by colorimetric titrations, or by several recently developed procedures. The electrode, when properly calibrated, gives direct readings of fluoride concentrations in test solutions. The colorimetric procedures can be of two types: those which show a color the intensity of which is proportional to fluoride concentration and those in which fluoride bleaches a chemical "lake" (a reagent dye preparation). Other procedures are based upon fluorimetry and special electrochemical techniques.

* Not a metal, but included in this discussion rather than in the series.

Approximate limit of detectability (by fluorescence quenching, but this degree of sensitivity seldom necessary)		0.02	µg
Air standard --	industrial (TLV) solid inorganic fluorides	2.5	mg/m^3
" " --	industrial (TLV) hydrogen fluoride	3.0	ppm
" " --	industrial (TLV) fluorine gas	0.1	ppm

Potable water standard variable with temperature:
Average annual maximum

F.	C.		
50.0 - 53.7	10.0 - 11.9	1.7	mg/li
53.8 - 58.3	12.0 - 14.4	1.5	"
58.4 - 63.8	14.5 - 17.4	1.3	"
63.9 - 70.6	17.5 - 21.4	1.2	"
70.7 - 79.2	21.5 - 26.4	1.0	"
79.3 - 90.5	26.5 - 32.5	0.8	"

```
Blood level  -- normal        0.01 - 0.03   mg/100g
Urinary level -- normal        0.2  - 1.5    mg/1 l
   "      "     -- excessive          4.0    mg/1 l
```

LEAD -- Atomic absorption spectrophotometric and colorimetric procedures are satisfactory. The latter (dithizone) method is being superseded in many industrial and clinical laboratories by the less complicated AAS technique. The amount of sample preparation necessary prior to assay is under current discussion, but some preparation is required in any case. X-ray fluorescence techniques are under examination for the estimation of lead in solids such as paint chips, painted surfaces, or filter deposits.

```
Approximate limit of
   detectability                      0.1     μg
Air standard -- industrial
             (TLV)           200.0    μg/m³
Water standard                        0.05    ppm
Blood level  -- normal              <40.0     μg/100g
   "      "   -- average             25.0     μg/100g
   "      "   -- excessive   50.0 - 80.0      μg/100g
```

Subclinical evidence of lead effect may be detected in this range.
Clinical plumbism may occur when blood lead levels are in excess of 80 μg/100 g.

```
Urinary level -- normal              0.03    mg/1 l
                                   (<0.06    mg/1 l)
    "       "    -- excessive
              (indus-
               trial       0.12 - 0.15      mg/1 l
```

Note: In cases of industrial organic lead exposure, very high levels of urinary lead may be excreted without evidence of clinical effects.

MANGANESE -- Atomic absorption spectrophotometric and colorimetric procedures are satisfactory.

```
Approximate limit of
    detectability                          0.05   μg
Air standard -- industrial
            (TLV)                          5.0    mg/m³
Water standard                            0.05   ppm
Blood level -- normal           7.9 -  28.0   μg/100g
Urinary level -- normal         7.8 -  18.0   μg/11
```

MERCURY -- The standard laboratory technique is usually based upon colorimetry. Field and other direct reading instruments are based upon the principles of atomic absorption.

```
Approximate limit of
    detectability                          0.1    μg
Air standard -- industrial
            (TLV)                          0.1    mg/m³
Water standard (tentative)                0.005  ppm
Human-tolerance level
    (mercury in fish)
    (tentative)                            0.5    ppm
Blood level -- normal                      5.0    μg/100g
Urinary level -- normal                   <25.0   μg/11
    "        "     -- occupa-
                    tional
                    absorp-
                    tion          10.0- 250.0    μg/11
    "        "     -- poten-
                    tially
                    hazardous
                    exposure             >250.0   μg/11
```

Note: a) Organic mercury compounds which are volatile, such as the alkyl mercury compounds, are readily determined by gas chromatography. Rapid advances in atomic absorption and neutron activation techniques for all mercury compounds are currently being reported. Levels of sensitivity in the 0.2 ppb range have been described.

b) Special note should be made of the fact that mercury and many organic mercury compounds are volatile and may be lost from improperly stored samples. In certain stored materials the conversion of non-volatile compounds to those which are volatile may lead to serious errors of analysis and interpretation.

c) There is a wide range in the toxicity of the several groups of mercury compounds. The present standards do not always recognize this fact, and the subject is currently under active discussion. Permissible levels of mercury in air or water or food may vary according to the types of compounds which are present. With standards based upon the nature of the compound, the analytical techniques used will have to provide more critical determinations than total "mercury". Criteria for mercury in water may not be based on the fact that soluble inorganic mercury compounds are hazardous at those levels but rather that various biological systems may transform and/or concentrate this mercury to create a potentially hazardous situation.

NICKEL -- Atomic absorption spectrophotometric and colorimetric procedures are very satisfactory.

Approximate limit of detectability	0.1	μg
Air standard -- industrial (TLV) nickel carbonyl	0.001	ppm
other forms	1.0	mg/m^3
Blood level -- normal	1.0 - 40.0 (avg. 4.2)	μg/100g
Urinary level -- normal	1.0 - 80.0 (avg. 10)	μg/1l

VANADIUM -- Atomic absorption spectrophotometric and colorimetric procedures are satisfactory.

Approximate limit of detectability	0.1	µg
Air standard -- industrial (TLV)	0.5	mg/m^3
for fume	0.1	mg/m^3
Water standard	--	
Blood level -- normal	1.0 - 10.0 (avg. 4.7)	µg/100g
Urinary level -- normal	2.0 - 42.0 (avg. 7.0)	µg/1l